# NUCLEAR CRISIS



Dr. Montefiore : 01-672-6697
Proliferation pp. 39, 43, 65/6, 142, 150

Published 1977 by:

PRISM PRESS
Stable Court,
Chalmington,
Dorchester, Dorset DT2 OHB.



Printed by:

UNWIN BROTHERS LIMITED,
The Gresham Press,
Old Woking,
Surrey.

0 904727 57 2 Cloth
0 904727 58 0 Paperback

# NUCLEAR CRISIS
## A Question of Breeding

*edited by*

Dr. David Gosling
*and*
Dr. Hugh Montefiore

PRISM PRESS
1977

# Introduction

This book contains an edited account of the Submissions and the cross-examination of Expert Witnesses at the Public Hearings on the projected Commercial Fast Reactor CFR-1 which were held at the London International Press Centre on 13 and 14 December 1976.

The method adopted was to invite national bodies and certain Expert Witnesses to submit 500 word statements on particular aspects of the matter. Some twelve submissions were received from British Universities. A distinguished panel of people with expertise in various aspects of nuclear energy agreed to serve; and they cross-examined witnesses on their submissions for a total of ten hours. Members of the public were invited to participate by sending written questions to the Chairman which were put to witnesses at his discretion. Naturally, the various members of the panel hold their own views on nuclear energy—that is evident from the form and nature of their individual questions. But the object of the Hearings was not to reach a verdict, but to expose the salient facts and the underlying principles, so that the ordinary person, reading what took place, may make his own informal judgement on this vital question.

D.G.
H.M.

# FOREWORD

This book is to be seen as part of a world-wide effort by the Churches to respond to the challenge of new technology, and especially nuclear technology. The World Council of Churches has promoted a series of events, begining with the Conference on Church and Society, Geneva, 1966. In June 1975, at Sigtuna, Sweden, the World Council of Churches convened a hearing on nuclear energy; see *Facing up to Nuclear Power*, edited by John Francis and Paul Abrecht, The St Andrew Press, 1976, £1.50.

The Church of Scotland Home Board initiated in 1970 the Society, Religion and Technology Project, directed first by Dr. John Francis and latterly by Dr. Colin Pritchard. The General Assembly in 1976 approved a deliverance from the Church and Nation Committee requesting the British Council of Churches to convene the public hearings now reported.

The Council invited the Rt. Rev. Hugh Montefiore, Suffragan Bishop of Kingston, and Dr. David Gosling, formerly a nuclear physicist, now in the Department of Theology in the University of Hull, to arrange the hearings, timed for 1977. After an interview with the Secretary of State for Energy, it became clear that hearings, to be of use, must be held in 1976 with only five weeks to make arrangements. No praise is too high for the energy and imagination shown by Bishop Montefiore and Dr. Gosling. They were greatly encouraged by the almost unanimous support offered by very many people.

The Secretary of State for Energy, Mr. Tony Benn, was enthusiastic in welcoming the hearings and participating. Sir Alan Cottrell, by his advice and support, greatly assisted the planning of the hearings. The members of the Panel and the expert witnesses co-operated with patience and good humour to expose the questions at issue, much relevant information and conflicting points of view. The officers of the Council are deeply grateful to all who planned and took part in the event, but especially the Chairman and Secretary.

The Council commends the book to its member churches and the general public as a reliable guide to the debate about future policy on nuclear energy. The Council has not, so far, committed itself to a judgement on this issue. On this occasion, it has sought to do publicly what is regularly done in its working parties at one stage of their work. The officers will be glad to receive advice on the usefulness of holding similar hearings on other matters.

HARRY MORTON
General Secretary

January 1977.

# Members of the Panel

PETER ADAMS
National Officer, E.E.T.P.U.
Fuel and Power Industries Committee, T.U.C.
Vice Chairman of Negotiating Committee with U.K.A.E.A.

Sir ALAN COTTRELL, F.R.S.
Master of Jesus College Cambridge
formerly Chief Scientific Adviser to the Government

Professor DAVID HENDERSON
Professor of Political Economy, University College, London

ALAN McKNIGHT
formerly Inspector General, I.A.E.A.

Dr. HUGH MONTEFIORE (Chairman)
Bishop of Kingston-upon-Thames

Lord KEARTON, F.R.S.
Chairman, British National Oil Corporation

KEITH RICHARDSON
Industrial Correspondent 'The Sunday Times'

GRAHAM SEARLE
Director, Earth Resources Research, Ltd.

Revd. Professor JAMES WHYTE
Professor of Practical Theology and Christian Ethics, St Andrew's University

# List of Expert Witnesses

Rt Hon Tony Benn
*Secretary of State for Energy*

Dr. D. G. Avery
*Deputy Managing Director, British Nuclear Fuels Ltd.*

R. Belgrave
British Petroleum *(appearing in private capacity)*

C. W. Blumfield
*Director,* Dounreay Experimental Establishment, U.K.A.E.A.

Dr. P. Chapman
*Director,* Energy Research Group, Open University

F. Cripps
Institute of Applied Economics, Cambridge University, *Advisor to Mr Benn*

Dr. J. Davoll
*Director,* Conservation Society

Professor Sir Richard Doll, F.R.S.
*Regius Professor of Medicine, Oxford University*

Dr N. Dombey
*Senior Fellow,* Science Policy Research Unit, Sussex University

F. R. Farmer
*Safety Adviser to the Authority, U.K.A.E.A.*

Professor I. Fells
*Professor of Energy Conversion,* Newcastle University

Sir Montague Finniston, F.R.S.
*formerly Chairman,* British Steel Corporation

Dr. N. L. Franklin
*Chairman and Managing Director,* Nuclear Power Company

Professor J. H. Fremlin
*Professor of Applied Radioactivity,* Birmingham University

M. Garner
*formerly Under Secretary,* Elect. Div. Dept. Trade & Industry

L. Grainger
National Coal Board *(appearing in private capacity)*

R. Grove-White
Council for Protection of Rural England

Sir John Hill
*Chairman,* U.K.A.E.A.

B. Johnson
International Institute of Environment and Development

Dr J. McVeigh
*Head of Dept of Mechanical and Production Engineering,* Brighton Polytechnic: *Vice-Chairman,* U.K. Section, International Solar Energy Society

Dr K. Mellanby
*formerly Director of* Monkswood Nature Conservancy

Dr R. L. Nicholson
U.K.A.E.A.

W. Patterson
Friends of the Earth

Dr R. Pease
*Director,* Culham Laboratory, U.K.A.E.A.

Dr J. Ravetz
Dept of Philosophy, Leeds University; *Secretary,* Council for Science and Society

R. N. Razzell
Central Electricity Generating Board

Dr L. E. Roberts
*Director,* Atomic Energy Research Establishment, Harwell

F. A. Romig
International Institute for Environment and Development

Professor J. Rotblat
*Professor of Physics,* St Bartholomew's Hospital Medical School

Dr L. Rotherham
Advisory Council on Energy Conservation; *formerly Vice Chancellor of* Bath university

S. Salter
School of Engineering Science, Edinburgh University

J. Surrey
*Reader in Experimental Physics,* Science Policy Research Unit, Sussex University

P. H. W. Wolff
*Deputy Director,* Nuclear Power Company

# Contents

# *Chapter I*

# Mr Benn's Statement and Replies

## MR BENN'S STATEMENT AND REPLIES

*The Secretary of State for Energy, the Right Hon. Tony Benn, spoke as follows:*

I would like to welcome the decision of the British Council of Churches to hold these Hearings which are in a way parallel with hearings held by the Select Committee and to some extent follow the discussions that took place at the National Energy Conference and at earlier public hearings on Windscale that occurred earlier this year.

The matter of the Fast Breeder Reactor is a very important and extremely difficult issue. The public debate now taking place of which this Hearing is a part, is a genuine debate in that Government has not yet taken a decision and therefore these Hearings will be in time to influence our thinking. I don't think I need go into the case for the fast breeder reactor in any detail, save to say that there are few experts in the world who doubt that a world energy shortage could develop on such a scale as to provide real limits to the capacity of humanity to meet its energy needs and that this could have very serious consequences for many countries in the world, perhaps less for us because of our other energy resources. But we are part of a world energy structure. All fuels have some uncertainty about them; oil because of political uncertainties; gas because of the uncertainty of how long it will last; coal because the question of persuading miners to go down and dig coal over a very long period ahead is bound to be in mind; and then in the case of nuclear policy generally and particularly the Fast Breeder Reactor, most of the anxiety has centred around safety. Here I should perhaps put the record straight and invite the panel to have a proper sense of proportion. The amount of skill that is available in the nuclear industry both on the scientific and the engineering side is such as to give us every confidence that the tasks that they undertake will be carried out with the very highest safety standards. And indeed compared to the number of lives lost over other fuels, the mines being very much in mind, the record of the nuclear industry is an outstanding one. But of course anxiety about safety rests not upon past performance but upon the fact that the time-scale of the hazards goes so far beyond human experience that when we consider how to approach these decisions it becomes a somewhat awesome responsibility. On environmental matters that lie beyond immediate safety, we have, of course, the benefit of the Royal Commission and I have no doubt that the recommendations that they made and the comments they made will be much in mind.

May I now turn to the way in which we are seeking to assess the factors that should shape our decision? The first one must necessarily be in relation to energy policy as a whole. At the moment there is no common agreement between the fuel industries in this country about the pattern of demand and supply and how they will interact over the next 20-25 years. A great deal of work has been done and one of the objects of the Energy Conference and the Energy Commission that I hope shortly to set up will be to harmonise the forecast and to provide us with as good and agreed an estimate as can be achieved. But even within the forecast that we finally agree as being the best and the most reliable there is bound to be a discussion about the best way in which this demand may be met and about the balance of demand for certain fuels, between coal and gas and oil and nuclear and conservation, and indeed non-conventional sources of energy. These choices lie at the heart of energy policy decision-making and make it a complicated but an important element in reaching the Government's decision about the role of nuclear power and the fast breeder in particular.

There are also international implications here which I may, perhaps, just touch on. One is the obvious advantages in a project of this complexity of getting the maximum international collaboration, not only in terms of cost but in terms of sharing experience and optimising the safety arrangements that might flow from a fast breeder programme. In nuclear policy generally, there are other factors, namely the desire to minimise the risks of the proliferation of nuclear weapons or nuclear materials. It could be argued that these point towards the desirability of this work being undertaken in very advanced societies that are politically stable. There is another set of factors to which reference has been made in public debate: I would describe them as domestic political factors arising out of two considerations. One is the problem of security and the risk of terrorism and the second arises from what happens when you have policies so complex that the democratic process finds it hard to come to terms with the choices that have to be made.

Certainly as a Minister with these reponsibilities now on and off since 1966 when I first became Minister of Technology, I have always found nuclear policy the most difficult because Ministers are not experts, they are not scientists, they are not engineers, they are not qualified to assess in any way the technical decisions that have to be made. And yet, whether you look at it in terms of the environment or safety or energy policy, or the massive public expenditure involved in all projects of this scale, it is essential that nuclear policy should be preserved within the democratic framework of control and not sub-contracted off to those whose only claim to reaching decisions might rest upon their technical qualifications. I think it would be very frightening indeed if we were to say that our fuel policy required us to adopt a technique of production like nuclear power which in its turn required the decisions to be taken from the process of Government answerable to Parliament and the public, and put into the hands of those whose special qualifications for deciding them would rest upon their technical knowledge.

The Ministerial role must be to be an interrogator. He is there to ask questions on behalf of the public of those who deal in the matters of nuclear policy. There are two views of a Minister's rôle. One is that he is a temporary manager of the operation which he oversees and the other is that he is a public representative in the Department whose task it is to see that all the essential choices are brought out clearly so that the public may know what is involved. In my opinion, particularly in this area, that is the role that a Minister should adopt.

May I make one other case for publishing all the information that is known to Government and the nuclear industry? If information is not published, a Minister is necessarily in the hands of those who have that information anyway and he does not have the opportunity to get comments from those who would be able to comment constructively if the relevant facts were available to them. Therefore, in pursuing a policy of open Government which I have tried to do as best I can, I have acted not only because I believed and believe that the public is entitled to know what is at issue but also so that by that process I have access to advice that is directed to the very problems about which I have to decide. It is for that reason that I have published the questions to the Nuclear Inspectorate and, when I met them this afternoon they told me that shortly they aim to publish their replies (I think just after Christmas) and that will provide further information for the debate to continue.

Finally may I just say this in conclusion? In the decision about the Fast Breeder Reactor, it seems to me that the most important thing is that we should take adequate time before we decide what to do. Some years ago, just after the oil price increases were announced, there was a worldwide crash programme of very rapid nuclear power station ordering. The situation looks slightly different today because, as everybody knows, the oil price increase led to, or was followed by, or accompanied by, a major economic slump in the Western world. We have the time now to take this decision steadily, and I believe that time has a great advantage for us in that we are not pressed to reach decisions before we have had a little bit more experience and before we have clarified some of the safety matters which are bound to be in the forefront of the public's mind. And therefore, if I have one point to contribute—at any rate as far as my own thinking is concerned before the Government reaches a view—it is that an issue of this magnitude must have time and there must be information available and this Panel today and these Hearings will, I hope, help people to assess the information and to be able to contribute their views constructively.

*Mr. Tony Benn answered questions as follows:*

COTTRELL You say we have the time before we have to take this decision about the Fast Breeder Reactor. Do you feel that there is enough time between now and say the end of the century because clearly when the reactor is ordered it will be some years before it is constructed and then there will be some years of running it and testing it, and only then can a decision be taken to build any more. So we shall be very late on in the century before anything more than the first of these will be available; and at that stage the predictions are that we shall be running low in hydrocarbons and fossil fuels, so that we may have a very difficult situation for energy in this country by the year 2000. Do you feel that there is time to fit in all these things?

BENN I am not suggesting, in the comment that I have made, that there is no pressure whatever, because of course there is. All the industries, all the fuel industries work on the same principle, namely, that they would like to meet the energy gaps themselves. Gas would like to fill it, coal would like to fill it, oil might be able to fill it, conservation argues that it could fill it, and clearly not everybody can fill it, and therefore you have to decide what is this energy gap, when will it come and what is a rational use of public investment in order to fill it and who should fill it and to what extent. If you are saying to me: is it now agreed that there is no alternative to the Fast Breeder Reactor? my answer must necessarily be that it is not agreed. Indeed if it was agreed that there has to be one, public debate would be meaningless because the decision, in a sense, would already have been taken. Obviously if the nuclear option is to be maintained, and I have myself little doubt that nuclear energy will play a very important part in our meeting of fuel needs, then there has to be an industry there in order to provide it; and that, I think, is a factor that is bound to enter into consideration in exactly the same way as the miners will argue that you can't close the pits and then expect to depend on coal later if the pits are not there to provide it. But these are factors which we have got enough time to consider with reasonable speed and I don't think we are under the same pressures that we may have thought we were a year or two ago.

CHAIR We are covering a very great deal of ground today in these Hearings. We shall hear much evidence. But it would help us very much if you could tell us something in the way of policy rather than practical details.

BENN Well, all decisions involving such levels of investment are bound to be political decisions because the amount of money involved is such that no Government could give a blank cheque to any particular solution to our energy problems. If you take alternative energy sources, a great deal of work is going on in this area, but I have in this respect to take advice. In order to prepare an adequate programme you have first to analyse the possibilities. You build up very slowly before you can get to a programme that can be fully backed; and then you have to look at the economics of it. The Severn Barrage for example, has been often mentioned and being a member for Bristol and also a Minister of Power on more than one occasion, my life has been divided between writing to the Minister asking for the Severn Barrage and replying saying that it is not yet economic! And that has been going on over a very long period. If you take conservation, conservation is undoubtedly capable of yielding a very considerable gain in meeting energy needs, but we haven't yet got a clear way of assessing the speed with which it could meet those needs or the cost or indeed the method of doing it. Since the decisions are entirely decentralised, Government can only do so much by changing the building regulations, by spending more money on insulating buildings that it builds itself, by encouraging industry and so on. There has been a fall in energy consumption, but we don't yet know how far that is due to the slump and how far it is due to wise energy management.

Personally I would like to see energy conservation given a much higher priority because the return on the capital invested is much more rapid. But it is expensive and one musn't think one can get energy anyway without public investment and public expenditure.

CHAIR Do you look for an increasing growth rate, with more energy required?

BENN Well, to predict the rate of growth of the industrialised countries at this moment over the next 5 or 10 years would be a very difficult thing to do. I think there are two questions in people's minds which go well beyond the energy problem. Firstly, will the rate of economic growth of the OECD countries continue at the rate we've seen over the last 25 years? Nobody quite knows that except that we are in the middle of a very serious slump.

Secondly, are we sure that energy consumption in relation to GDP will continue at its past relationship? Now you can predict as best you can but no man could confidently say what is the answer to either of those questions. That is why energy policy in my judgement can only be hammered out in public by those with some contribution to make, on an annual basis, updating one's assessment as the situation becomes clearer and develops.

ADAMS On the question of time, do you believe that we have time if we are hoping to co-operate with European countries on the development of Fast Breeder Reactors, bearing in mind that some people think that we have left it too late anyhow?

BENN Well, I have been involved in many international projects in my time and the argument has always been used that if you don't go very quickly, you are going to miss the boat. My own view is that France and Germany, which have entirely different energy patterns, are being driven by their own problems to a much bigger nuclear programme on a more rapid scale than would make sense to the United Kingdom; and the theory that we must be in the forefront in every single fuel at the same time doesn't seem to me self-evidently to be the case.

SEARLE Energy is often substituted for manpower, while at the same time we have a problem with large numbers of people unemployed. Do you yourself see a shift back to a greater dependence upon a different form of energy—our own metabolic energy—rather than a continuing substitution of that by fuel, hydrocarbons or nuclear energy,?

BENN We have lived for so long in an economic system where, if you lay people off, this is held to be synonymous with productivity, that we have forgotten the fact that we must provide employment for people, and there is no doubt whatever that conservation, insulating homes, building flues, etc, is very labour intensive; and just as the Americans came out with the Environmental Impact Statement, they are now considering Employment Impact Statements to compare the amount of employment generated by alternative ways of meeting particular problems. Although I wouldn't want to think that this is just job creation in a limited sense, if you have a choice and there is an equal gain in terms of social expenditure on either solution and one creates employment and the other does not, it may very well be that advanced industrial countries now confronting technological and structural unemployment on a really big scale with no immediate answer yet available to us, might find that Employment Impact Statements would be relevant and would point or incline towards conservation against the generation or development of new power sources.

CHAIR (Question from the floor). Does the Secretary of State regard it as possible that the CFR-1 if adopted could ever be admitted as unworkable or unnecessary after so much money had been spent on it?

BENN The one thing we don't lack in this country is critics of our technological performance. I think I could rephrase the question differently. If the Government decided to build CFR-1 can you think of any moment when it would ever have a favourable press from the moment it started until the moment it came out of service?

McKNIGHT What is the attitude of Government towards the danger of proliferating nuclear weapons and nuclear weapon technology?

BENN Oh, totally opposed; and one of the problems (and it is in a way a moral problem) to which the Panel might wish to give its mind is this. Britain with its record of political stability and social responsibility and expertise in nuclear power is quite clearly from a world point of view a much better place to do a lot of nuclear activity than some other countries, and it has been seriously argued that if there is to be, as there must be, reprocessing, because there is already nuclear power in use quite apart from future decisions, it would be better that it be done in this country rather than a spread of reprocessing technology with its necessary generation of plutonium in other countries; and the thought that we might have, in some way, a moral and political responsibility to develop nuclear power within the framework of our own society rather than have it go elsewhere is a factor which has to be taken into account.

# *Chapter II*

# Future Needs and Goals for Society

The Hearings began with questions about future energy requirements at a time when conventional sources of energy such as oil and uranium will be likely to become more and more scarce. Witnesses were questioned about the possibility of additional energy sources becoming available in sufficient time and in sufficient quantity and appropriate quality, to meet this future demand. Some consideration was given to the social consequences of having and of not having what many people regard as the most promising additional source of energy, the Fast Breeder Reactor (F.B.R.). Attention was given to the implications for life-styles of future energy supplies

## Projected Energy Needs and Conservation

The first two submissions by Mr Robert Belgrave of British Petroleum (appearing in a private capacity) and Dr Peter Chapman, Director of the Energy Research Group at the Open University, gave varying accounts of the quantity of end-use energy which is likely to be needed by the year AD 2000. Both stressed the difficulties involved in establishing valid criteria for projecting from the current situation into the future, the former arguing mainly in terms of economic growth rates, the latter suggesting that domestic consumer "saturation" will reduce future levels of energy demand.

*Mr Belgrave (B.P.) made a personal submission as follows:*

Over the 25 years from 1948-1973, energy consumption in the industrial countries grew at around 4½% a year, that is to say it doubled every 15 years. The rate of growth of energy consumption in the developing countries was over 6½%. Over the years, the increase in energy consumption has matched the increase in economic activity, that is to say a 1% increase in economic activity has led to a 1% increase in energy consumption. Whilst it is possible to believe that this relationship could be altered by more economical use of energy, it is hard to believe that the ratio could be altered by more than a third and then only if there is general government and public acceptance of the need for severe conservation measures and appropriate pricing. In that case each 1% economic growth would require 0.66% increase in energy.

Government policy in the industrialised countries presumes at least the historic rate of economic growth. Therefore, if national styles of life, including adherence to economic growth and full employment, are not to be drastically changed, energy demand in industrialised countries is likely to be at least double the present level by the end of the century.

It makes no sense to consider the U.K. in isolation. Even if it was desirable or politically possible to isolate ourselves from the rest of the world, the option of a fortress Britain does not exist. North Sea oil and gas will give this country energy self-sufficiency for some years from about 1980, but production from existing and future discoveries is expected to begin to decline some time after 1990 or by the end of the century this country will once again have to import oil, gas, coal and uranium.

The non-communist world depends on oil for 50% of its energy. Professional opinion is that new discoveries will shortly cease to keep pace with annual consumption, and that some time after 1990, world production will begin to decline, though it will not suddenly cease. By the year 2000, the remaining oil and natural gas will increasingly be used for transport and chemical feedstock, including fertilisers, for which alternatives are least suitable. Unless there is to be a rapid decline in industrial activity, other sources of energy must be developed by that time to fuel industry and to provide heat. Only coal and existing proven nuclear designs can make any significant contribution to the gap in this time frame, but these themselves may not be sufficient to fill the gap after 2000. A decision now not to proceed with the development of the Fast Breeder Reactor would therefore leave us short of energy in the 21st century.

Mr Belgrave was questioned as follows:

SEARLE What is "economic growth" and is it true that Government policy in countries like our own presumes at least a historic rate of growth?

BELGRAVE Every Government in all industrialised countries assumes and portrays its attachment to some economic growth however defined. You can define it in terms of gross national product, you can define it in terms of material goods. But whatever the conventional definition that you use it is inevitably associated with consumption of energy; and if you increase economic growth and you increase employment, as has been the experience in the last 25 years, some increase in energy consumption has to go with them.

COTTRELL Would you give us your picture of the distribution of fuel consumption for this country at the end of the century? How would you see it proportioned between oil, coal, nuclear, gas etc?

BELGRAVE The United Kingdom consumes about 100 million tons of oil a year, or will by 1980, and this will rise from now until then. It consumes about 130 m. tons of coal (which is equivalent to something less than 100 m. tons of oil in terms of the energy which is produced), and one assumes that that will be maintained. The U.K. also uses one third of that quantity of natural gas. Some of that is turned into electricity, but nonetheless electricity has got to be generated at the moment from coal or oil or gas, and a minute quantity from nuclear sources. Those are the main components. Towards the end of the century we hardly know how that will change. If oil is available one would suppose that since it's an efficient and convenient fuel the proportion might be about the same. But it seems doubtful whether this country will be producing enough oil to meet those requirements which one has to assume will have grown by then, and so it is a question of whether oil is available for import.

COTTRELL Do you then imply that there will be a considerable demand for coal and nuclear by the end of the century?

BELGRAVE If we are to increase our economic activity I think that is so.

WHYTE The idea that there should be no drastic changes of life style was wished upon Mr Belgrave by the form of the question. But considering the great changes that have occurred in the last 25 years which are associated with an increased use of energy, would you not say that our life style has been, in fact, changing radically over the last 25 years, and that what you are looking forward to in terms of a greatly increased use of energy is a continuing change in life style?

BELGRAVE Yes, I agreed with that. I think the implication of what you are saying is that we could use energy more economically as time goes on, but I would stress the long lead time, the inertia of change.

RICHARDSON You say that energy demands in industrialised countries are likely to be at least double the present level by the end of the century. What account have you taken of the fact that the relative cost of energy has gone up a great deal and is expected to continue on almost any presupposition about technology? The capital cost of providing energy is going to be so great that they will want to devote an increasing proportion of their income to financing new energy sources, or is it conceivable that their demand will drift on to other things?

BELGRAVE I am not saying that anything is likely or unlikely, I am saying that if you believe that there will be an increase in economic activity that is going to require more energy. Because of the cost and because of the shortage, I think people will be more economical in its use, and I hope that they will. But I do not think that the relationship between increasing economic activity and increasing energy use can be changed overnight. It is a gradual process.

CHAIR Have you taken into account the likely shortage of resources and the difficulty, say, of extracting metal from ore when you say more energy will be required?
BELGRAVE Yes.

*Dr Chapman, Director of the Energy Research Group of the Open University, made the following submission on the same subject:*

Future energy needs cannot be reliably gauged from an examination of historical trends since these trends were generated by a combination of exceptional circumstances which are very unlikely to recur.

1. Over the past 50 years successively easier sources of energy have been exploited and over the period 1950-73 the real price of fuels was declining. The decline in real prices gave a weak incentive for using fuels efficiently. Energy prices are likely to rise in the future since all the suggested sources of energy are significantly more expensive (in terms of capital, labour and resources) than natural gas and oil.

2. Since 1900 the decline in the number of people per household (from 5 to 2.75) has led to a rapid increase in the number of households which are the significant consuming units. The number of people per household cannot halve again. The future growth in consuming units is expected to be close to the rate of population growth, which is low.

3. Over the past 50 years there have also been very large improvements in personal mobility and levels of thermal comfort in buildings. However, with 0.75 cars/household and half of all dwellings centrally heated these trends are now approaching saturation. This approach to saturation will significantly slow the rate of growth of fuel consumption in these areas, which together account for almost 40% of total fuel use.

4. The rate of acquisition of many consumer durables, such as fridges, cookers, TVs, washing machines, etc. is also slowing as ownership of these approaches 100%. This will stabilise the demand for these industrial products and the materials used to make them. It is most likely that future economic growth will be concentrated in automatic control systems (e.g. in cars, heating systems) and communications equipment (e.g. Ceefax; videophone; etc.) These economic sectors have a much smaller energy requirement per pound value added so that even if economic growth continues at 3% per annum the growth in energy demand (in industry) will be significantly less than the historical rate.

5. The approaching saturation in the domestic sector, the likely shifts in the industrial sector and the higher future price of fuels encouraging more efficient use, lead to a significantly smaller future rate of increase in energy demand. Our detailed calculations suggest 400 to 450 million tons of coal equivalent (m.t.c.e.) per year by 2000 and 600-700 m.t.c.e./year by 2050.

Dr Chapman responded to questions as follows:
McKNIGHT Is not the saturation to which you refer a cause of

unemployment in this country?

CHAPMAN It would be much too simple to say that unemployment is caused by any one or two single factors. But I certainly think that the change in the trends of motor car ownership and ownership of central heating appliances and so on are contributory factors to the depth of economic recession we are in.

SEARLE We've heard that it is likely that energy demand will double from its present level by the end of the century. Your predictions are that by the end of the century we will need 400-450 metric tons of coal equivalent per year, which is nothing like a doubling; would you like to comment on the last assertion about the doubling being inevitable?

CHAPMAN There can't be any inevitability about any prediction or look into the future. The link between economic activity and energy is not held at an aggregate level in the sense that the energy required, to produce a pound of added value, is very different between different products. So if we have a G.N.P. growth which is fuelled by increased production of fertilisers and steel, then the energy required will be very much larger than if we get the same economic growth fuelled by an increase in the production of cosmetics and services. So it is not possible to say there is a definitive and unalterable relationship between economic growth and energy consumption.

SEARLE Your forecast is that energy demand will double by the year 2050. Are you then agreeing on the same future scenario as Mr Belgrave and the same future technical problems and just differing about the time scale? Or do you really think that a high energy use society can be avoided altogether?

CHAPMAN No, I think it is a question of time scale that is at issue, and I would say that you can't belittle it. If you really do have an extra 50 years in which to make up your mind about what you do on energy policy, that really is important. So although at the end of the day the last speaker and myself would certainly be agreed that this country is likely to get up to an energy consumption of 600-700 m.t.c.e. per annum, it does matter when you get there.

SEARLE What difference does it make?

CHAPMAN It makes a lot of difference about the time you have to develop a whole series of energy saving and energy producing technologies.

SEARLE To what degree do your forecasts reflect the type of future that you would like to see as opposed to what will occur?

CHAPMAN I recognise that all forecasters cannot keep their own prejudices from entering their work. However our forecasts are low because of saturation effects. For example, fifty per cent of houses in this country now have a central heating system. The growth in central heating systems will therefore start to decline because once every house is centrally heated, you reach a saturation point. Now that is one factor which is very important. Another important factor is that we are past the turning point in the ownership of motor cars and their utilisation. I am not suggesting that people will stop using motor cars, but they will use them less. The rate of increase of use of motor cars is going to slow up. That seems to me to be inevitable when I look at the saturation curves which indicate that when every person in the country has got a car and they are driving it four or five hours a day, that's saturation. Now you might say, O.K. let's switch to flying aeroplanes. But then you would be talking about a different type of life style which we haven't considered. We are not trying to say that this is caused by price. We are saying that it is caused by coming to the end of the road of a particular development of a life-style. What comes after it we wouldn't like to guess.

COTTRELL You've projected that we shall need about 50% more energy per year at the end of the century than we use now and yet we have been told that it's unlikely that the amount of coal that will be produced will be greatly increased and that the oil will be running down at the end of the century. If you agree with that, then what do you see as filling the gap between your projection and what can be supplied?

CHAPMAN Rather than try to predict this, I would urge you to consider it as a matter of policy. If you foresee a difference between the sorts of energy supply that would arise without taking any action and the sorts of energy supply that you require, then you are faced with a question of policy; and policies developed over the next 25 years could increase the output of the British Coal Industry. Policies could control the depletion of oil and gas reserves offshore, so as to maintain quite a sensible output in the year 2000. I have given what I regard as the sort of potential outputs of the various industries which could fill the gap in the year 2000.

WHYTE What do you see beyond our present life style?

CHAPMAN Almost infinite variety. You take your choice. There are a number of areas of high technology developments which seem to me very sensible—such as increased communication systems and improved control systems and so on. There is also the possibility of reducing the length of working time and of making an economy which is more directed towards the satisfaction of consumers as labourers rather than towards production for productions's sake. It would take at least 50 years to bring about some of the substantive changes in life style which would be sufficient to alter our energy consumption pattern. Therefore, we should make sure that any decision or policy

recommendation that goes forward doesn't prejudice any life styles. It seems to me wrong to prejudice or presuppose some life style change for a policy that we are trying to implement today.
CHAIR If we kept our energy requirements at the present level would you see a decrease in life style for the population of this country by the year 2000?
CHAPMAN You are using the word decrease, you could have used better or worse, you could have used any one of these very value-loaded words. The problem is that what you consider better, I might consider worse. It seems to me very important to try to distinguish those things which correspond to our needs, among which I would include a warm house and enough food, from what we regard as activities of leisure, enjoyment or further enrichment. I do not see any way in which a group of people in a society which has some sort of democratic procedures would actually lead itself on a task where collectively everybody was agreed they were worse off. It seems to me that by the definition of a democratic society, we will always be getting better off.

Sir William Hawthorne's submission on behalf of the Advisory Council on Energy Conservation on future energy requirements presented the anticipated situation of the U.K. against a background of world energy supplies between now and the year A.D. 2000, and considered four projections corresponding to high and low economic growth coupled with alternative assumptions of high and low indigenous energy supplies. In all four cases there would need to be a net import requirement for coal, oil, and gas by the year A.D. 2000:

> All four projections indicate a rapid transition in the U.K. from an energy export potential in 1990 to an energy import situation in 2000. This transition is expected to take place against a world background in which energy supplies are increasing too slowly to match the earlier trends in demand.

The submission proceeded to indicate areas in which the U.K. Government could play an important role, and called for rigorous measures for energy conservation:

> The U.K. Government can influence a number of areas that relate to future energy prospects, including:
> (i) The future development of the coal supply industry.
> (ii) The development of more efficient techniques for the direct utilisation of coal so that it can compete more effectively with oil and gas for a larger share of the market.
> (iii) The rate of development and depletion of North Sea oil and gas.
> (iv) The future development of the electricity supply industry, particularly nuclear power.

(v) Energy conservation measures, particularly those arising from regulations and standards, and the pricing of different energy supplies.

(vi) International energy policy, including energy pricing and the co-ordinated international response to a possible future scarcity of energy supplies.

For the short to medium term future U.K. energy prospects are favourable compared to those of many industrialised countries. These favourable prospects could lead on to energy problems in the long term unless there is adequate development of the coal and nuclear power industries *together with the adoption of vigorous measures for energy conservation*. These developments involve long lead times and they require sustained programmes during the medium term future, when the U.K. may be a net exporter of energy. *Additional measures for energy conservation in this period will contribute also to the longer term future by allowing more time for the development of alternative supplies and for long term gains from energy conservation.*

CHAIR I would like to ask you some questions arising from the 1976 Review of the International Energy Agency, *Energy Conservation*. We are told that United Kingdom utilities experience a very low utility capacity factor and conversion efficiency. What scope exists for conservation measures to be taken in this country which will really reduce the amount of energy which we require?

ROTHERHAM An improved load factor within the Generating Industry doesn't necessarily mean that you have an economy. The amount of coal used is equated with the amount of units sent out and not in any way with the load factor utilisation of the equipment. There would be some gain but the fact is that the stations which are used during the night are the more efficient stations, so you produce more electricity for given coal consumption when the load factor is low than when it is high.

CHAIR We are also told that in most of the specific efficiencies for industry, the United Kingdom ranks the lowest average in I.E.A. countries and that a non-subsidised loan programme has disbursed only £90,000 in 1975. Have you any comment?

ROTHERHAM As far as the efficiency of utilisation is concerned, it does not matter what you use. For example, the utilisation of fuel in diesel engines in this country is not very different from that in other countries. The load factor in power stations may very well be lower but this is a straightforward matter concerning when the capital was invested. Stations which have been built more recently are more efficient than those which were built a long time ago. Capital expenditure and actual running costs are all related.

GRAHAM SEARLE Could you not stop persuading people to use more energy?
ROTHERHAM I didn't say you couldn't persuade people to use less energy. The Conservation Council's duty is to try to persuade people to reduce their energy, and in some measures I believe it is succeeding. The rate of growth that is postulated for energy usage in this country is reduced if you make reasonable assumptions on what can be done with Energy Conservation. But I don't think you can eliminate a growth in energy demand.
SEARLE Would you say that people in Britain are persuaded by advertising and competition between different sorts of energy providers to use more energy than they should be persuaded to use? Is that not a positive wastage?
ROTHERHAM We have suggested that the nationalised industries could be persuaded not to advertise competitively and thereby increase energy consumption. They are all represented on the Council and I think they all realise that this is a policy that in the not so distant future they will be compelled to follow wh ether they like it or not.
RICHARDSON If a customer is prepared to pay the full economic cost of a given slice of energy because he thinks it brings value to him, what is the case for the Government trying to lean on him and either to prevent him from burning that energy or to tax him for doing so?
ROTHERHAM There is no case for imposing energy conservation and disregarding all the consequences of that action, unless you are willing to make assumptions of what our life style will be and what industrial production will be at the end of the century. I don't think this is a thing that can be done at the present time.
CHAIR Could you spell out a little what is the meaning of your phrase "the adoption of vigorous measures for energy conservation"?
ROTHERHAM We have divided the work of our Council into the work of a number of sub-committees. If you take the industry committee, the immediate action which has been taken is to provide advice and help to industry, essentially to make economies which can be achieved quickly by good housekeeping. A lot has been done and I daresay it is quite fair to say that considerable economies have been made. We have heard that about 50% of energy is used in housing. The Department of the Environment, as you are probably aware, are looking at building regulations to see what can be done to economise on the use of energy in housing. We have a transport sub-committee which has examined the use of transport and come forward with a series of recommendations which have been put to the Department of Energy and to the Department of the Environment. The

committee of which I am a member, the Publicity and Education Committee, arranges meetings with teachers and others so that the facts can be brought into education at an early stage.

A further estimate of energy needs in the year 2000 was given by Professor Fells, who maintained that 525 m.t.c.e. of energy will be required, and whose contribution will be considered in detail in the next section. But it will be apparent that projected energy needs in the year 2000, with or without conservation, are subject to a large number of variable factors which make them extremely difficult to calculate. At one end of the spectrum Mr Belgrave, while declining to give a precise figure, believes that "energy demand in industrialised countries is likely to be at least double the present level by the end of the century", whereas Dr Chapman's estimate, based largely upon an argument that the domestic market for consumer goods is becoming increasingly saturated, is considerably less (50% more than present demand, i.e. between 400 and 450 m.t.c.e.) Opinions differ on the time-scales involved, with the Advisory Council on Energy Conservation predicting a major transition from an energy export potential in 1990 to an energy import situation in 2000, while Dr Chapman argues that shifts in energy strategy can be incorporated into a much longer time-scale. But whatever the time-scale all agreed that ultimately much more energy will be needed and that the question of additional energy sources will have to be pursued in earnest. A diminished life style (in terms of G.N.P.) for the U.K. and the Western world was not considered.

**Additional Energy Sources**

Detailed consideration of alternative sources of energy will be given in Chapter V. Here the prospects for energy increases of the order suggested above are considered, and it will continue to be assumed both that conservation measures will be applied and that here will be no drastic changes in life-style. Granted these assumptions, will the exploitation or harnessing of new sources of energy become necessary by A.D. 2000 to meet anticipated energy requirments?

Professor Ian Fells of Newcastle University answers the question in the light of his own estimate of a 525 m.t.c.e. demand for energy in A.D. 2000. He argues that even if the energy gap can still be filled by conventional methods by the end of this century, the situation will have changed drastically by the year A.D. 2010. His submission was as follows:

There was a reduction of about ten per cent in the annual consumption of primary energy in the U.K. between the traumatic happenings of autumn 1973 and the end of 1975. However figures available for 1976 show a reversal of this trend and during the third quarter total inland energy consumption rose by four per cent over the equivalent period in 1975. Taking into account the warm summers of 1975 and 1976 and the general recession in industry during this period it is probably unreasonable to suppose that energy consumption will remain static. Predictions of growth in energy demand or for that matter of gross domestic product (the two are linked but not as closely as some people suppose) over a period of twenty-five years are fraught with difficulty. Nevertheless I propose to choose a fairly conservative growth rate in energy demand of two per cent per annum and look at the ways in which we might expect to satisfy our anticipated demand in 2000 A.D.

Total annual demand for primary fuel by 2000 A.D. will on this projection be around 525 m.t.c.e. (million tons of coal equivalent). Table I shows generally acceptable figures and also optimistic figures for primary energy demand in the U.K. At the present time the electricity generation industry accounts for one third of this primary energy but, because of losses in generation and transmission, it provides only one eighth of the energy used by final consumers. It follows that a gradual switch to a more electricity orientated energy economy over the next twenty-five years will result in higher primary energy demand figures than those quoted for coal and nuclear energy inputs. Oil accounts for nearly half of end use energy, natural gas a fifth, electricity an eighth and solid fuels of various kinds the rest. The pattern of end use energy is probably rather inflexible if life and industrial styles are not to change significantly. However oil usage will probably be in decline post 2000 A.D. and its place in the energy mix will be taken up by electricity. The oil generated heat load will be replaced by electricity, leaving oil for transport and petrochemicals. Part of the heat load will also be taken up by waste heat recovery, alternative sources and improved efficiency of energy usage. Using generally acceptable figures for what will be achievable by 2000 A.D. and assuming a substantial (35 m.t.c.e.) contribution by alternative sources of energy (wave, tidal, solar power, etc.) the "acceptable" projection still calls for a large nuclear component. However the total energy production still falls short of the predicted demand which could be made up by coal or oil imports or a bigger nuclear component, (although this might be difficult to achieve in the time scale).

If one puts into the projection optimistically high production rates for coal, oil, gas and alternative sources one might possibly achieve the target figure of 525 m.t.c.e. for 2000 A.D. without calling upon a nuclear component. The realisation of these optimistic figures would require very high investment in manpower, materials and money (starving other needy parts of the economy) and if one takes into account technological difficulties the chances of achieving all of them must be slender. Should such a course of action be pursued the real problems arise post 2000 A.D. as oil and gas production from the North Sea and elsewhere decline. Taking the year 2010 A.D. primary energy demand on a two per cent projection will have risen to 640 m.t.c.e. (i.e. double in thirty-five years).

Using the following production assessments for primary energy

| | 2010 A.D. |
|---|---|
| Coal | 200 |
| Oil | 100 |
| Gas | 50 |
| Alternatives | 90 |
| Total | 440 m.t.c.e. |

this leaves a "gap" of 200 m.t.c.e. which could either not be filled or filled with nuclear power.

The U.K. is fortunate in having indigenous supplies of oil and gas as well as coal and they just might enable the U.K. to pursue a low or even zero growth nuclear programme up to 2000 A.D. The foregoing argument suggests however that this policy would be difficult to sustain post 2000 A.D. The "energy gap" in other E.E.C. countries will arise much earlier and can only partly be filled with imported oil or coal. Even the nuclear solution to filling the gap raises the problem of uranium availability although this might be solved by "breeding" reactor technology.

It seems therefore that the only ways of avoiding a growing dependance on nuclear power in the U.K. and maintaining a growth in energy supply which in turn means a steady if slow improvement in life style is either through a doubling of the coal industry or a gigantic increase in alternative sources or both. The prospect of increasing the coal supply seems to be limited to a maximum of thirty per cent above the 150 m.t.c.e. figure and in the E.E.C. the situation is much less encouraging.

What are the prospects therefore for a very large increase in energy supply from "new sources of energy"?

The fusion reactor gives promise of almost unlimited power with only a moderate radiation hazard. The reactor has not yet been properly controlled in the laboratory so that a "fusion" power station must be at least thirty and probably fifty years away.

Other possible sources of energy are based on "income" or "renewable" supplies and include solar, geothermal, tidal, wave and wind power. Of these solar energy and wave power have the greatest potential. The solar energy input to the U.K. is about eighty times our current requirements and wave power with a potential of 70 kW. per metre length of wave (at 100 per cent recovery) and an enormous gathering area of wind generated waves from the seas surrounding the U.K. also looks attractive. Geothermal energy once extracted takes a long time to regenerate, tidal energy would be concentrated on the Severn Estuary yielding a useful but limited 6 gW (maximum) of generating capacity and wind power is only appropriate in rural locations. All have or would have very considerable impact on the environment and require heavy financial investment. The chief limit to the energy obtainable from these attractive energy sources is the rate at which the embryo technologies can be developed on a large scale. This is not helped in the U.K. by our current predilection for spending ninety per cent of the energy R and D money on nuclear power and relying on other countries to develop alternative technologies, notably the U.S.A. Bearing these constraints in mind a figure for cumulative energy supply from all these sources of 45 m.t.c.e. by 2000 A.D. is optimistic. This could only be increased by a revolutionary change in our energy investment and more particularly our priorities.

In the U.K. other possible sources include oil shale, the potential of which is still unquantified and energy obtainable from waste material which is not recovered at the present time.

It should be emphasised that the U.K.'s "energy portfolio" is different from and in some ways superior to those of the other members of the E.E.C. but it would be dangerous to follow an isolationist energy plan as co-operative ventures will be essential because of the enormous investment necessary if either nuclear or alternative strategies are to be developed.

*Table I Generally expected and optimistic figures for primary energy demand in the U.K. (million tons of coal equivalent)*

| | *2000 A.D. expected* | *2000 A.D. optimistic* | *2000 A.D. measured* |
|---|---|---|---|
| Coal | 150 | 200 | 118 |
| Oil | 175 | 210 | 134 |
| Gas | 60 | 75 | 55 |
| Alternative Sources | 35 | 45 | 2 |
| Nuclear | 50 | 100 | 11 |
| Total | 470 | 630 | 320 |
| Total excluding nuclear | 420 | 530 | 309 |

Professor Fells was questioned as follows:

RICHARDSON You say that the cumulative energy supply from all new sources of energy could be 45 m.t.c.e. by the end of this century. But you say such a figure is optimistic. Could you give us what you would regard as a realistic figure?

FELLS It is very difficult to do that because it depends on how much money we are prepared to put into these enterprises rather than into the feasibility studies that we are so keen on funding. A realistic figure could be 25 m.t.c.e. a little lower than the figures proposed by the Marshall Plan.

RICHARDSON When a question is put to the Government and the Electricity Boards, they always say that there are no projects yet in sight which would justify putting larger sums of money into them. Is it your contention that there are such investments within reach?

FELLS I would have thought that the Severn Barrage is a scheme which could be built with today's technology. Yet the kind of assessments which are made every five years or so keep on suggesting that the Severn Barrage is not feasible. That would give you 6,000 megawatts of electricity. We keep on being told that wave power is the best bet, and of course when you look at it, it does seem attractive. But speaking cautiously as an engineer, it is a problem of scaling things up. What seems attractive on the laboratory bench turns out to be very difficult when you try to build it kilometres long. Even the Atomic Energy Authority finds it difficult to scale up from a 250 megawatt reactor to a 1300 megawatt reactor. You have got to build things with a certain amount of faith in the hope that you can then build something rather bigger.

COTTRELL You've said that to meet the U.K.'s growing energy needs after the year 2000 we shall have to depend either on nuclear power or on a doubling of coal or on a gigantic increase in an alternative source. You've just talked about the alternative sources. Do you feel that a doubling of the coal industry is a practical possibility?

FELLS I asked Sir Derek Ezra two weeks ago what the prospects were, and he thought they were fairly bleak. He said that they might get up from 150 to 200 m. tons of coal a year. The prospect for doubling it would mean radical re-appraisal of the coal industry, and the environment would almost certainly suffer under those circumstances.

CHAIR Would that include the subterranean gassification of coal?

CHAIR The underground gassification of coal is one of those technologies that we tried in the late 1950's and it didn't work. In fact the gas that was produced wouldn't burn. You had to add propane to it to make it burn, so it wasn't a very attractive proposition. To abandon that technology and to say that since it didn't work 20 years ago, it won't work now, is a mistake. We need to work more at it, though it is a long way away.

CHAIR In your paper, you suggest that we shall require by the year 2000 about 525 m.t.c.e., whereas Dr Chapman suggests that we shall need between 400 and 450 m.t.c.e. I realise that there is a wide bracket here and that predictions are very uncertain but the difference between these two is quite considerable. Would you care to comment?

FELLS I am certainly not wedded to 525 m.t.c.e. In my preamble I said if we take a pessimistic growth rate of 2%, it comes out at about 525 m.t.c.e. You can manipulate it a little. Certainly I have no desire to be pinned down to that sort of figure. I am prepared to believe that we can settle for a lower figure than that, if we use our energy more sensibly and as long as the energy conservation techniques are taken seriously instead of letting everything go on as before.

*The Central Electricity Generating Board submitted the following on the future needs and goals of society:*

1. The Central Electricity Generating Board's statutory function is to develop and maintain an efficient, co-ordinated and economical system of supply of electricity in bulk for all parts of England and Wales. It is also required to take into account the effect which its proposals might have on the environment.

2. At present the Board has in operation 3½ Gw of Magnox nuclear plant. The first of the four Advanced Gas-Cooled Reactor stations with a total capacity of about 5 Gw is now coming into operation. The only nuclear project at present planned by the Board is the 2½ Gw Sizewell 'B' station based on the Steam Generating Heavy Water Reactor (S.G.H.W.R.) system chosen by the Government in 1974 for the next nuclear programme.

FUTURE NEEDS

3. Because of reduced demand for electricity, there is no need to order any new generating plant of any description before about 1980. The Board's current view on electricity demand were set out in its 1976 Corporate Plan. For the period up to 1990, two energy scenes are postulated, the first assuming a moderate electricity growth rate of 3.4% p.a. and calling for a new plant installation rate of 1-2 Gw p.a. in the mid 1980's, rising to 5 Gw p.a., requiring no additional new plant beyond Sizewell 'B' until 1990. It seems likely that actual events will lie between these two suggested cases.

4. For the longer term, there have been numerous predictions about future fossil fuel availability and the possibility of a world shortage developing. When this will occur is debatable but few people doubt that such a shortage will occur. In the United Kingdom there is general agreement that an energy gap could occur in the 1990's which it will only be possible to fill by a combination of coal and nuclear energy. Nuclear power will have to play its part if a drastic reduction in living standards is to be avoided. The present public debate has produced no indication that reduced living standards, or change in our way of life, would be welcomed by the broad mass of the population.

5. The main justification for fast reactors is that they are capable of utilising uranium resources, which are also limited, far more effectively than thermal reactors. They are therefore a major consideration in any national strategy aimed at ensuring adequate supplies of energy for the future. The possibility of a shortfall in energy resources occuring around

the turn of the century just cannot be ignored, and to do nothing would risk mortgaging the future.
6. It is only through the electricity system that the fast reactor could make a contribution to the country's energy requirements. The Board's own particular interest in the fast reactor is therefore as a potential purchaser of the system, albeit in the long term.

Mr Razzell kindly came forward from the audience to answer questions in connection with the C.E.G.B.
COTTRELL What is the Board's view of the cost effectiveness of the nuclear power that you have been using?
RAZZELL The Board are lucky, in that we bought our nuclear power stations sometime ago. They have, in fact, paid for themselves in terms of the capital involved and the operating costs. They are cost effective, bearing in mind that they were paid for in pound notes some long time ago and we are paying the 'mortgage', if you like, in current pound notes which aren't worth so much.
McKNIGHT How does your cost of electricity vary with the cost of uranium? If there is a price rise in the cost of uranium, will that affect your electricity bill?
RAZZELL Yes, of course. I am afraid I don't know the exact breakdown of costs on nuclear power stations, but if for example, the cost of uranium were to go up by a factor of 1,000, then nuclear power would be very uncompetitive with other methods of producing electricity, in today's terms.

So far the assuptions of a present life style and type of society had gone unquestioned. *Dr John Davoll, however, Director of the Conservation Society, made the following submission:*

1. The terms of reference of the questions misdirect the inquiry by apparently confusing *meeting the real needs of society* with *preserving a particular set of institutions and schemes purporting to meet these needs*. By identifying "life style" with the latter, and excluding "drastic changes" in it, a meaningful descussion of "Future Needs and Goals of Society" is unjustifiably precluded.

2. Thus, we may assume that the following components of our current life style are to be left substantially intact:

(a) The overall aim of national policy, to which other considerations are to be subordinated, will the maximization of a crude aggregate represented by a single scalar number—G.N.P.—and containing, *inter alia*, the costs of car production, traffic jams, accidents, factory production lines, mental hospitals, polluting industries, pollution controlling industries, etc., etc.

(b) An economy oriented towards the maximization of the rate of flow of material rather than the maintenance and increase of stock.

(c) A transport and land planning system designed to provide mobility for the minority, rather than access to facilities for the majority.

(d) A continuing attempt to evade social and ethical problems by real or promised increases of material production.

(e) Relative indifference to any factors in people's lives that cannot be reduced to monetary terms.

(f) A greed for energy that makes acceptable the sale of arms and nuclear facilities to the most politically unstable parts of the world.

3. Much of this life style developed during a period when both energy and the raw materials to whose manipulation it was to be applied were cheap and expected to get cheaper. Even in those favourable circumstances its ability to meet real human needs efficiently was open to doubt, and now that cheap energy has come to an end, and the extravagant use of raw materials is (or should be) unacceptable, even more stringent questioning is appropriate. Ways should be sought to meet the country's needs in ways that fit the new circumstances, and are also compatible with reducing the gap between the rich and the poor countries.

4. Nevertheless, institutions such the U.K. Atomic Energy Authority whose corporate aims necessitate open-ended expansion are naturally slow to recognise the new need for qualitative changes in society as opposed to undifferentiated expansion. The underlying general philosophy of expansion and unlimited expectations has made endemic in our life style inflation, increasing social tentions, environmental degradation, and lack of any sense of common purpose. Is this what we wish assiduously to preserve from drastic change?

5. We also consider that plans for the deployment of the fast breeder economy are open to criticism on more conventional grounds, and hope that these will be covered by other contributors to this Inquiry. However, we believe that an open-minded consideration of life styles different from conventional projections should be anterior to detailed descussion of specific topics.

Dr Davoll answered questions as follow:

KEARTON In your paper you cast doubt on whether our present standard of living is necessarily a good thing, and even more doubt on whether improvement of our present way of living would be a good thing. Do you seriously subscribe to that point of view?

DAVOLL I mean what I said.

KEARTON Am I to take it that you are advocating a reduction in the consumption of energy and therefore a reduction in the material standard of living as generally accepted by most people?

DAVOLL I think one has to be a little more specific here. If one belongs to the minority of the population that has the discretionary use of a car, then what I propose might reduce what they consider their standard of living. But if you belong to the majority that doesn't have the discretionary use of a car, the pedestrians, the view I put forward would increase the pleasantness of their lives. One cannot give a general answer to the question you asked. One has to discriminate between different sectors of the population.

KEARTON You refer in your paper to real human needs without defining them. Perhaps you would define them?

DAVOLL They were defined, I think, by Dr Chapman. He mentioned that food, warmth and shelter are necessary. Obviously we have other needs. We like to get to places where there are facilities that we wish to enjoy, but whether we do that in a car or whether the land planning is so arranged that we can walk or cycle to them, seems to me not very important. In fact the advantage is that we should be able to walk or cycle to them, particularly if we wish to drink as well!

KEARTON I think that the car is responsible for 10% or so of our energy use and space heating in this country in one form or another in the home and factory and so on represents about 40% of our energy use. As I understand it you are not terribly enthusiastic about cars. Do you still regard space heating as being a good thing?

DAVOLL Yes, and I would suggest that what we are trying to find is a comfortable temperature to live in, and we should achieve that not by pouring in more heat to keep the sparrows warm on the roof but by insulating the house. The energy for heating can be supplied by small district heating schemes, by combined heat and power systems of the sort that don't seem to have achieved any popularity for reasons connected with the built-in corporate aims of the power generating industry.

McKNIGHT I appreciate the numerous references throughout the papers before us to the use of waste heat, but this seems to me to be a terribly capital intensive way of using heat, particularly with our system of siting nuclear power stations in fairly isolated situations away from residential areas. Have you examined that question?

DAVOLL If you build large power stations according to policy, it costs a lot of money to use the waste heat, but I was suggesting that we should go over gradually from this policy towards a smaller size power station from which the waste heat can be used more economically than at present. I realise that many of the things I am suggesting take a long time to carry out, but this is an argument for starting them.

McKNIGHT There was a proposal many years ago for putting glass houses on the sites of nuclear stations and using the heat to grow food. Have you ever examined that question?

DAVOLL It might or might not be advantageous. It seems to me to be one of those marginal projects which might or might not be worthwhile. I don't think it is one of the central questions. The more important is a systematic policy designed to meet a whole variety of needs within a given expenditure of power.

McKNIGHT Do you often encounter a lack of support in connection with building regulations and local government policies for the conservation of energy?

DAVOLL One might say that the main trouble here is that, in choosing how to make a building, the emphasis is all too often on minimising capital costs rather than on paying regard to running costs. If people had to take account of running costs over a period, then, especially at a time when energy costs are rising, we would see a much more rational approach to building design.

RICHARDSON Obviously, from your statement, you dislike a very large number of features of the society in which we live. You hit out at cars and a very large number of other things. It is not clear to me whether you think that your dislike is shared by a majority of people in this country. What institutional changes will be needed to shift the life style in the direction you want? Do you think this will come about by democratic means? Do you think that they will have to be imposed on people? I am thinking particularly about suggestions that people should not be allowed to use cars, that people should have to use housing with a higher capital cost, that perhaps people should go in the direction of central heating schemes such as district heating with complicated cost factors involved. Would people voluntarily vote for such a society?

DAVOLL There are two points to make. If you ask people whether they are in favour of cars, they have either got cars or in many cases hope to get them. So they will say yes, they want cars, because people are not used to taking into account the cumulative effects of their own actions being generalised in society. Thus you might find that the same people who say yes, they want cars, also complain that their children are in danger on the roads when getting to school. But they don't perceive the link between their individual pursuit of advantage and the collective degradation of society. This is a well known problem of social organisation. One cannot, of course, easily compel people to do things, but demonstration schemes can be set up so that people will gradually perceive that restrictions in the use of cars at which they might jib at first, do produce improvements in their surroundings. I would agree that the people who are better off are less likely to perceive this because they can usually afford to opt out of the consequences of their use of cars. This is a political matter. Secondly, will people accept this voluntarily? The answer is probably no, because decisions in society are made by well-off people who own cars and are quite prepared to inflict the disamenities of using the car on others while enjoying its advantages. As I see it, changes will be enforced by circumstance. It seems to me that the Panel makes very widespread sweeping assumptions that our future is under our control. I think, in fact, that circumstances will force us in directions which we don't regard as acceptable. We may not want to accept it, but by God we will have to do. In these circumstances we can move in two ways. Either we can move into a polarisation of society in which the rich grab what they can and the poor sit there—the Chilean solution—or alternatively, and more constructively, we can move towards an increase in communal facilities so that the reduction of private affluence is counterbalanced by improvements of public welfare. But I agree that this is a political matter and we cannot say how it will be resolved one way or the other.

WHYTE Dr Davoll mentioned in his paper the gap between the rich and poor countries. I would like to ask him how he sees the different alternatives that are before us in relation to their effect on this gap?

DAVOLL One alternative is that we should accept some reduction in our own material wealth in order to pay higher prices for the goods we get from the poorer countries. The other alternative is to use our undoubted ability not to do this, and face the consequences. If I had more time I would tell you how unpleasant they might be.

**Possible Implications of Fast Breeder Reactors**

Having considered possible energy requirements at the end of the century and beyond, and the extent to which additional energy sources may need to be needed, social implications which might follow as a result both of using and not using Fast Reactors were next considered.

Statements by the C.B.I. and the T.U.C. on the subject of Fast Breeder Reactors were noted. The C.B.I. had, in its submission to the National Energy Conference, noted an awakening concern in industry regarding the prospect for energy supplies in the late 1980s and beyond. This concern extends to

> The need, again concurrently, to plan for the eventual conservation of diminishing reserves of fossil fuels through better techniques for their utilisation, and for their substitution initially by nuclear power pending the development of other substitutes or supplements from new energy sources and forms. In this connection, an early review of the current nuclear power programme with a view to its eventual extension would seem to be necessary;
>
> Finally, but no less importantly, the need to provide substantially for a large and wide-ranging R & D effort towards new and 'benign' energy forms. A start has been made on this with the recent grant for research into wave energy, but this and other deserving forms of research will need to be increasingly heavily backed. In this connection, the possibilities for international co-operation and mutual assistance through the strengthen-

ing of E.E.C. and I.E.A. relationships should not be decried.

The T.U.C., in a submission for the same Conference, stated:

> An area of delay could be associated with the reactor to follow the S.G.H.W.R. The T.U.C. recognises that the role which will be required of the Fast Breeder Reactor (F.B.R.) in the late 1990s demands an intensification of the research effort.

Scarcely any of the witnesses volunteered opinions as to what might happen to society should there be insufficient energy available at the end of the century, though Mr. Grainger of the National Coal Board, submitting evidence in a private capacity and unfortunately unable to be present at the Hearings, claimed that undersupply of energy could create more social problems than efforts to expand the energy supply base. His submission underlines a number of important points which emerged during the first session of the Hearings.

> Society is likely to have more serious problems from the under supply of energy than from efforts to expand the energy supply base. Whether energy is expanded or not, the whole energy system will be almost completely revised by about A.D. 2000, mainly as a result of the scarcity and/or cost of gas and oil. This transformation can be accomplished in a way which is socially, economically and environmentally acceptable but much better standards and planning will be necessary. Without nuclear power however the problems of meeting public aspirations in living standards would be very difficult early in the next century, though not so crucial before then. . . .
>
> Key decisions on energy policy have in the past been made on inadequate grounds and logic. It is duty of the technologist to present clear alternatives in understandable terms to government and the public, especially where acceptibility and economics are linked.

There was no submission from Friends of the Earth but the following extract from *Non-nuclear Futures* by A. Lovins and J. H. Price was included, with their approval, as representative of their position. They were represented by Mr. Walter Patterson.

> In the few countries where the matter has been seriously studied, chiefly the U.S.A., it appears that even modest efforts at energy conservation—those that are economically worthwhile, use existing technology, and have no significant impact on lifestyles—would make nuclear power unnecessary; and even in the short term, conservation would certainly "provide breathing room so that we can gain a better understanding of nuclear power problems, and reach some better judgments before major new expansions of nuclear power are made." (The purely economic incentives favouring such conservation efforts are overwhelming). Indeed, on present knowledge it is not clear that there is any country that cannot by such means, augmented in many cases by a wide range of non-nuclear energy technologies, fulfil liberal economic and

social goals without further recourse to fission. But few countries have had much incentive to explore alternatives—including energy supply that is more decentralized, nonelectrical, and based on sources of energy income. Most such supply and conservation options are technically and socially simpler than fission technology; many seem to be cheaper; none has had even a hundredth of the R & D support accorded to fission, and it is thus not surprising that most are now less developed, whatever their technical merits may be. Until all alternatives to fission have been explored far more thoroughly, the argument for fission as a last resort (rather than as a first resort) seems premature. The basic question should be: if we devote our time, skill and money not to fission but to energy conservation and "soft" energy supply technologies, starting now, can we use fossil fuels wisely as a bridge to help us construct over the next few decades an economy of energy income, without an intervening stage of reliance on fission?

Faced with a choice between starting a transition to an economy of energy income now, when we have the fossil fuels to subsidise the transition, or later, when we shall not, we would do well to bear in mind the irreversibility of a commitment to nuclear power, a property unique in the evolving biological world that Geesaman and Abrahamson eloquently describe:

> Man has lived until now by the chemical sublimation of the energies of sunlight. His ancient evolution has been a biochemical process. Life is mediated by collective stabilities in a space dominated by chemical energies. Persistence stems from a biochemical wisdom that is far more pervasive than conscious thought. . . . We have lived consciously with nuclear energy for a generation. In that time, we have pondered it cautiously as if it were a sleeping giant. In our wars, we have been careful to leave it undisturbed. Now, when our society is fragmented and alienated, when it is threatened by a diminished stability and a fading sense of self-duplication, is it the proper time to try our hand at nuclear energy? In this enterprise, the old reserves of evolutionary stability may be useless; and we must persist "as on a darkling plain" by the working of infallible intelligence alone.[1]

CHAIR If energy consumption is this country became dependent on the large scale use of fast breeder reactors, would there be any implications good or bad for our industry, deployment of labour, environment, culture and security measures which you think would affect human happiness and public welfare?

PATTERSON The question is not a yes/no question on this issue, it is very definitely an either/or question. It is a question of whether we move in the nuclear direction particularly to the direction of fast reactors or whether we move in a number of other directions about which some of my colleagues and others at these hearings are actively involved in carrying out research and development. A move in the direction of fast reactor electrical technology as the main source of energy supply for the country presupposes an increasing centralisation of the decision-making process, and an increasing polarisation of the social system as a result. This is because eventually the decision making will have to be in the hands of a very few highly qualified people, and it seems likely to me that the employment pattern will also be greatly influenced as a result. For example, we now see "work productivity" used always to represent the substitution of capital and energy for human skills. This is undoubtedly one of the major contributors to the problem of increasing unemployment which we see not only in the United Kingdom but in many other industrialised countries.

RICHARDSON You suggested that you felt that fast breeder reactors were somehow a key step in the direction of a centralised highly polarised society in which the decisions would be taken by a very small number of very highly technically qualified people. Do you feel that since we've had 20 years of nuclear society, we have already advanced a long way in that direction? Do you feel that we are already in a state where key decisions are taken by a very small number of highly qualified people? Do you feel that these changes could be avoided in the future?

PATTERSON It isn't clear that they would be avoided. I don't myself think that we are as committed to a nuclear society as seems commonly to be the impression. At the moment, we are getting a contribution of about one per cent of our total energy from the nuclear sector. This means that we still have a great deal of room to manoeuvre on the variety of different options. My own choice would be to go not for large scale centralised systems (like the Severn Barrage) but for smaller systems including total energy systems which can be built where you want them and when you want them, rather than having to build power stations which you have to order five years before you know you are going to need them because of the time that it will take to build them.

COTTRELL What fuel is going to be used for these local stations?

PATTERSON My choice would be coal in the transitional period in the U.K. for the next 75 to 100 years.

COTTRELL But coal supplies less than half of this country's energy at present and the minimum forecast at these Hearings has been that the country will need 50% more energy by the end of the century. This means that coal will supply a third or less of this country's energy by the end of the century. How do your calculations fit?

PATTERSON The Science Policy Research Unit of the University of Sussex has carried out a study which demonstrated that since the late 1950s the per capita end-use energy which is used for heating, for space heating particularly and domestic water heating, has been more or less constant for close on 20 years. The only correlation which they were able to find between the increase in per capita primary energy use in the U.K. and obvious end use sectors, was the increase in the use of private transport. Therefore our problem, if we do have an energy gap in the year 2000 and thenceforth, is going to be for the supply of liquid fuel and I would suggest that the emphasis on nuclear electricity is likely to be largely irrelevant to that problem.

*The submission from the U.K. Atomic Energy Authority was addressed in general terms to all the issues which have been considered so far:*

The large scale use of fast reactors which could start in about 20 years' time when the development and proving programme will have been completed, would have the following advantages:

1. fossil fuels would no longer be required in such vast quantities for electricity generation and would be freed for roles for which there are no easy substitutes;
2. the utilisation of nuclear fuel would be improved about fifty fold—an unparalleled contribution to energy conservation; and
3. the environmental benefits of nuclear power—a clean, safe, non-polluting form of power production—would be available on a larger scale than now.

2. Problems associated with safety or social impact would not be significantly different for fast reactors than for thermal reactors for which, in turn, no insurmountable problems exist.

3. If fast reactor development should now be halted the consequences could be serious in the future. The energy resources available to the world community would be much reduced, with the result that:

(a) the economic activity which supports our community will have an additional constraint imposed upon further growth;

(b) it will be very much more difficult, if not impossible, to meet the legitimate expectations for a gradual improvement in the standard of living of the under-developed countries of the world.

4. To bring to an end rapidly the prospects of economic growth and—because of the need to use increasingly dilute and scarce natural resources—to risk a decline in material welfare would create a variety of social and economic problems. To minimise these problems one must plan for such changes if required to take place over a very long time-scale.

5. The following factors are relevant:

(a) economic growth per capita is largely brought about by the substitution of energy for other inputs;

(b) because of limits inherent in the technologies concerned, there is now only modest scope for continuing further improvements in the efficiency of energy conversion which, from the turn of the century to the beginning of the Second World War, permitted economic growth with little growth in energy consumption;

(c) there is no scope for continuing the progressive substitution of oil and gas which, since the end of the Second World War, permitted the U.K. to have a primary energy growth rate only two-thirds of its G.N.P. growth rate—compared with a ratio greater than one for the world as a whole;

(d) schemes such as combined heat and power systems offer some scope for further economies but the cost and practical limitations for such measures are often under-estimated: they are unlikely to play a major part over the next few decades;

(e) North Sea oil production will pass the peak in about 10 years' time and world oil production will peak in about 15 years' time, and the scope for expanding coal output on this time-scale is limited.

6. It follows that without a large new source of energy we would be unable to maintain standards. Of all the possible new sources, the use of uranium in fast reactors stands out as the only source that is large enough and mature enough to meet the situation. This is seen by most of the other developed countries, and it would seem an act of economic and social masochism for the U.K.—which still leads the world in fast reactor technology—to opt out of its exploitation. We advocate that the U.K. play its part in international collaboration to complete the development of this technology.

Sir John Hill, Chairman of U.K.A.E.A. represented the Authority:

SEARLE Would you like to defend the statement about 'large scale use of fast reactors which could start in about 20 years' time and . . . would have the following advantages'. I would like to hear your defence of 'the environmental benefits of nuclear power—a clean, safe, non-polluting form of power production'. Does that include the whole fuel cycle and storage of waste?

HILL Yes, it certainly does. The whole experience of nuclear power has been that of a clean and safe way of providing the energy we require. I see no reason why, with an expanded programme, we shouldn't continue to have this clean and safe form of energy. Whatever we do, we leave some problem to future generations. I believe the storage of a relatively few hundred or thousand tons of solidified waste is one of the least problems that we will leave to our grandchildren. The big problem is going to be the shortage of hydrocarbons for the next generation and the one after that. I am quite sure in my own mind that the storage of waste, provided it is run competently and sensibly, is not going to be any problem for future generations.

CHAIR The third point in your submission is that if fast reactor development should now be halted, 'the consequences could be serious in the future.' I was interested in the choice of the word 'could'. What are its implications? Do you mean that there are mitigating possibilities (such as conservation, increased use of coal and then perhaps fusion), or do you mean that the consequences are bound to be serious?

HILL I believe they would be serious, but we are talking of long term development here. We are not discussing powering Britain with fast reactors in the next five years or so, but a development programme where we have already built 20 years ago a 15 megawatt fast reactor, we then built a proto-type reactor of 250 megawatts, and we are now proposing that we should build one full-scale fast reactor power station which couldn't be in operation is less than 10 years from now. So we are really discussing what energy availability there will be in the world towards the end of the century. I think all the evidence we've heard is that the shortage of oil and gas towards the end of the century is going to result in a major world energy shortage. To deny ourselves this opportunity of having fast reactors at that particular point in time, would be most serious for the world as a whole.

WHYTE I believe sufficiently in original sin to think that once we have a very strong interest in the matter, we tend to believe what we want to believe. Is our judgement going to be a little clouded in its favour once we've put in such an investment? In other words is there not a kind of inertia so that once you start, you carry on?

HILL It qould be quite irresponsible to suggest that if we found that fast reactors were unsafe we would continue to go on building them just because there was some inertia in the system. I believe in taking measured, careful steps in the development of nuclear power. I am totally against having programmes in areas where we don't understand what we are doing. So the whole development of fast reactors in this country has been a progressive forward movement. When we understand one set of problems, we move forward another step, and what we are proposing now is the next step. Unless we thought there was a very high probability of the full scale fast reactor working completely satisfactorily, we wouldn't ask to go ahead. If this did come up of course the programme would be stopped.

COTTRELL One of the submissions at these Hearings suggests that nuclear energy cannot be a major supply of energy for us at the end of the century, implying that we have left it too late and that we shall hit the energy shortage because we will have used up hydro-carbons before neclear energy can take the strain of that. Is his your view?

HILL We've already left the development late. I am not a person who believes in crash programmes. I suspect the world will wake up to a very serious energy shortage in perhaps 10 years time. Then people will be clamouring for massive nuclear programmes. I am advocating now that we do the necessary preparatory work while we have time, so that we can then go ahead in full knowledge of all the implications of what we are doing.

A submission was made on behalf of the National Nuclear Corporation and the Nuclear Power Company and Dr Franklin, its Chairman and Managing Director, was questioned on it. Due to a technical oversight his cross-examination was not recorded. Since the points in this submission were all raised elsewhere during the Hearing it seems fair in the circumstances not to include it in this record.

---

[1]Lovins A. and Price J. H., non nuclear futures, (F.O.E. 1975) p. 46, 53.

# *Chapter III*

# Risks

## Technological risks of CFR-1

Since the Royal Commission on Environmental Pollution had only recently reported, three excerpts from their Sixth Report were considered instead of submissions:

(i) Possible accidents and hazards to personnel and to the community

*Fast Reactor Safety* The F.B.R. has the advantage that the liquid sodium coolant is not pressurised and has a high heat capacity. Whereas in thermal reactors the coolant has to be kept in motion mechanically by pumps, in the F.B.R., even if the pumps failed, natural convection in the sodium would maintain a high rate of heat removal and would allow a time of the order of an hour before auxiliary cooling would have to be brought into use. This would be supplied by one of several alternative systems provided in the design. On the other hand, there are problems in handling sodium on a large scale, since it constitutes a major fire hazard. Experience suggests that these problems can be solved and even quite large sodium fires have been satisfactorily extinguished. It should be noted that the presence of sodium precludes the ultimate use of water to extinguish a reactor fire, this method was used eventually to extinguish the fire in the Windscale reactor which occurred in 1957.

The use of a low pressure system in the F.B.R. avoids the problems of high-pressure containment which make it difficult to ensure the structural integrity of thermal reactors.

With any reactor employing a liquid coolant there is the risk that if a substantial amount of the fuel should melt it could react explosively with the coolant, causing the latter to vaporise. A reactor that has been running for a little while will contain significant amounts of gaseous fission products in the fuel (notable krypton and xenon) and these would tend to form a blanket to prevent intimate contact between the fuel and the sodium. Still, the possibility exists, and is particularly serious for the F.B.R. because of the risk that the explosive forces generated might so change the core geometry as to lead to a prompt critical situation. There would then be a cumulative increase in reactivity which would be limited only by the core blowing itself apart, that is, by "Disassembly" of the core, and its vaporisation and dispersal. No mechanism can be foreseen by which this could happen in gas-cooled Magnox and A.G.R. reactors though a possible mechanism can be postulated for water cooled reactors. In such reactors the increase in reactivity that could be caused by the rapid removal of a control rod might be sufficient to bring about prompt criticality; this would imply that the rod was blown out of the core but measures to prevent this possibility are included in the design. Thus the F.B.R. is not unique in posing the hazard that a prompt critical condition could arise, but the event would be much less improbable for the F.B.R. where a greater range of circumstances might lead to its occurrence.

Earlier in this chapter we discussed the effects of a major release of the gaseous and volatile components of the fission products from a thermal reactor. The inventory of fission products for an F.B.R. is likely to be similar for a reactor of about the same output capacity, but the core will contain several times as much plutonium, and an even greater proportion of the higher actinides, americium and curium. If fuel were to melt inside the sodium, most of the fission products would be retained within the sodium because of chemical reactions. If some burning sodium was expelled from the reactor, the fumes would carry any associated fission products. It is considered that for most such accidents that can be envisaged the fission products would be retained within the very strong structure within which the reactor is housed. However, the possibility of gross vaporisation of the fuel, and the subsequent escape of some proportion of the inventory to the atmosphere, overrides all other considerations. This would allow not only gaseous and volatile fission products to escape but also those that are normally non-volatile. It has been estimated that a 1 per cent release from a large F.B.R. would have consequences one to two orders of magnitude worse than those we have described for a thermal reactor. The major effect would be due to the inhalation of the non-volatile fission products, leading to the development of lung cancers, by people down wind of the release. Ground contamination would be caused by plutonium as well as caesium.

The consequences of such accidents would be so severe that the F.B.R. could scarcely be contemplated on an extensive commercial basis unless it is established with a very high degree of confidence that reactivity accidents can be controlled so as to prevent gross fuel vaporisation. It is generally considered that present research programmes will lead to substantially increased knowledge of these effects within the next five years. Work is proceeding in many countries as it is recognised that the problems are of fundamental importance if fast reactors are to be widely used. One important source of knowledge is operational experience from prototypes such as the P.F.R. at Dounreay. The behaviour of fuel under irradiation and under a variety of simulated emergency conditions can be studied, although the latter conflict somewhat with the need to establish reliability as a normally operating power plant. The incidence of local boiling of the sodium coolant must be detected quickly, and the means likely to be used is to monitor the system for the noise that such boiling would cause. There are, however, other safety issues that can be resolved only in the design process itself, and thus the design and construction of a first F.B.R. of commercial scale is an important step in assessing whether the required level of safety can be achieved.

### (ii) Disposal of Long-Life Toxic Residues

*Radioactive Waste Management* We have described briefly in Chapter III how radioactive wastes are generated at various stages of the nuclear fuel cycle, and in Chapter V we discussed the administrative arrangements that exist to control and authorise the discharge of some categories of waste to the environment. In this chapter we consider the subject of radioactive wastes in more detail. We shall be mainly concerned with those wastes whose management presents particularly difficult problems and which are at present stored at nuclear sites, and with the steps that are being taken to ensure that they will not cause undue harm to the environment.

In Chapter IV we stated our view that there should be no commitment to a large programme of nuclear fission power until it has been demonstrated beyond reasonable doubt that a method exists to ensure the safe containment of long-lived, highly radioactive waste for the indefinite future. These wastes already exist in substantial quantity and a safe method for their long-term disposal is in any case required whatever is decided about nuclear development in the future. We are clear that such a demonstration will require a substantial programme of research. In the latter part of this chapter we consider the possibilities that exist for the safe disposal of these wastes and the organisation needed to pursue the search and judge its results.

### (iii) Possible civil disorders or wars

We have given some thought to the possible effects of war so far as nuclear installations are concerned; these installations, providing vital energy supplies, would be prime targets. In a nuclear war the effects of attack on nuclear installations would be one part of a general catastrophe, but an attack with conventional weapons leading to the release of radioactivity would produce some of the effects of nuclear weapons. The quantities of fission products that could be released are vast and they would not be carried up into the stratosphere.

The effects of war, even of "conventional" war, are inevitably horrifying, but if these effects could be magnified by attack on nuclear installations, then this is a major factor to consider when deciding whether, or to what extent, to use nuclear power. This threat also exists, and should likewise be weighed, in the non-nuclear field. The vast increase in the chemical process industry over the last few decades has created many industrial plants where dangerous substances are used or stored and where the consequences of damage from armed attack could be extremely serious. The unique aspect of nuclear installations is that the effects of the radioactive contamination that could be caused are so long lasting. If nuclear power could have been developed earlier, and had it been in widespread use at the time of the last war, it is likely that some areas of central Europe would still be uninhabitable because of ground contamination by caesium.

Professor Sir Richard Doll, F.R.S. gave evidence on behalf of the Commission, and was cross examined by the Panel:

KEARTON Am I right in supposing that the Royal Commission felt that a trial CFR-1 is a sensible undertaking?

DOLL We certainly thought the preparations for the design of a trial CFR-1 would be sensible; but it would be only in working out the design that one could decide whether it was practicable to go on further.

RICHARDSON Is it possible that by drawing attention to the very high maximum credible accident, you are perhaps making it difficult for people to get a fair view of the probability of such accidents compared with lesser accidents which might be more probable?

DOLL You'll find that we didn't like the concept of the maximum credible accident. We preferred the concept of trying to work out the probability of different accidents of different degrees of severity.

SEARLE Do you think that before we expand in any way existing nuclear technologies and facilities we ought to find an answer to the long-term waste problem?

DOLL The Commission did not recommend stopping the current programme before this had been worked out but it did say that it shouldn't be expanded in any way until a satisfactory solution had been demonstrated.

COTTRELL Do you consider this disposal of waste a technical or an economic problem?

DOLL Our understanding is that it's essentially a technical problem.

ADAMS Are you saying that the question of disposal of waste and the design of a first fast breeder reactor should go together; that there is no objection to the fast breeder reactor design going ahead at the same time as investigations on the disposal of waste?

DOLL Construction could probably proceed at the same time but not actual use. We see no harm in construction going ahead as long as there has been ample public discussion of the issues before it is started. My colleagues had no doubt amongst themselves that there had not been ample discussion of the issues up to now.

WHYTE Considering that we're dealing here with a waste product with an active half-life of 24,000 years, do you feel there is

a dimension of responsibility here that one doesn't have with any other technological project?
DOLL Yes.
WHYTE You suggested that you had confidence that the problems would be solved. Is this a kind of general confidence in the ability of scientists to solve problems when they are identified, or is it grounded in the way things are going?
DOLL The evidence that we had on the waste disposal problem gave us no reason to doubt that if money was spent on appropriate research a solution could be obtained within 10 years.

*The following submission was received from the National Nuclear Corporation and the Nuclear Power Co. Ltd:*
Technological risks comprise:

(i) Possible accidents and hazards to personnel.
(ii) Possible accidents and hazards to the community.
(iii) Decontamination of outworn nuclear power stations.
(iv) Disposal of long-life toxic residues.
(v) Possible civil disorders or war.

This note presents the invited views of N.N.C. and its subsidiary the Nuclear Power Company Ltd, which is the U.K. organisation formed by the Government to undertake the design and construction of nuclear power stations.

Most large-scale industrial plants which use or produce energy have the potential to cause considerable damage to humans or property. Because of its recent emergence and military background the civilian use of nuclear power may fairly be claimed to have been subject to the strictest safety regulation. We accept this as appropriate for a new technology and an outstanding record for safe operation of power reactors has in fact been achieved. We also consider that in almost all respects the answers are not materially different for fast reactors than they would be for corresponding thermal reactors. Furthermore, we would not expect that operating licences would be sought or granted for fast reactors unless they could be demonstrated to be comparably safe with the thermal reactors which are their contemporaries.

The safety design of all types of nuclear power station is based on an "in depth" approach with the objective of ensuring to very low levels of uncertainty that an initial equipment failure, operational error or human intervention will not lead to severe consequences. The measures used include:

Design characteristics chosen with tolerance to fault conditions as a prime consideration.

Protection systems to detect faults, if they arise, and retain the plant in a safe condition. (To ensure reliability independent triplicate or quadruplicate sets of equipment are used.)

Safety systems to control the consequences of failure of all the above protection systems (e.g. the reactor and its auxiliaries are enclosed in a strong, leak-tight containment building).

All types of reactor of a given power output contain similar amounts of fission products and plutonium. It is necessary to be assured that the above measures can be applied equally effectively to a fast reactor with its particular characteristics as they can and are to thermal reactors. The objective will be to ensure that at the end of a highly improbable accident sequence the reactor contents remain within the containment. If this condition is met the hazard to personnel or community is virtually zero. If it is believed not to be met to any significant extent, permission to build or operate such reactors would not be sought or granted. If the belief that the requirements has been met proves false through human ignorance the consequential injury or death is not likely to be on a massive scale.

Considerable experience has been obtained so far with experimental and part-scale fast reactors in a number of countries. The purpose of constructing a small number of full-scale demonstration fast reactors in Western Europe is to increase that experience and to demonstrate the option to embark upon a programme of fast reactor construction in the 1990's if such a course should seem necessary. We envisage that the phasing of the construction of such a reactor in the U.K. would be an integral part of this purpose.

Decontamination of stations at the end of their life has to be considered for any type of reactor. Some advantage is expected for fast reactors because of the compactness of those components which could become highly active and which are designed for easy removeability in many cases. The induced activity in the primary sodium decays relatively rapidly; in principle it could be re-used for new plants.

The disposal of long-life toxic residues arising from the operation of fast reactors is not significantly different from that required for thermal reactors because the residues are similar. We believe that the Sixth Report of the Royal Commission on Environmental Pollution gives a balanced account of the problem and of the necessary demonstration work. Sufficient is already known to give us the highest confidence that such work can be made successful: the primary responsibility for its pursuit is at present with A.E.A./B.N.F.L.

For a given electrical programme (i.e. number and size of plants) we do not consider that fast reactors are more vulnerable to damage from acts of war, nor are the potential radiological consequences of such attacks very different from those for thermal reactor stations. We also expect any power station type to require similar safeguards against civil disorders and we consider them to be less sensitive in this regard than a substantial number of other industrial installations.

Dr Franklin , Chairman and Managing Director of the Nuclear Power Co. gave evidence:

KEARTON I wonder whether you could tell us what has been the experience of the Dounreay Fast Reactor in regard to safety of

operation?
FRANKLIN I don't know of any reasons which would give rise to any anxieties, but it is, I think, fair to point out that the Dounreay Fast Reactor was operated with enriched uranium fuel in the main and not with plutonium.
CHAIR What about the Phoenix Reactor? Does what's happening in France give you confidence for us?
FRANKLIN I believe so. I think that the difficulty in making assertions about confidence is that it has several different aspects. If one's talking about reliability of operation that's one thing, but if one's talking about horrifying accidents whose probability is generally admitted to be very very small indeed then to that extent operating several reactors for several tens of years doesn't really provide full statistical evidence.
RICHARDSON It's commonly said that fast reactors pose safety problems of an order of magnitude greater than thermal reactors particularly as regards a nuclear explosion that could arise in certain situations. Do you agree?
FRANKLIN One would have a potentially different problem in relation to the largest size of explosion that one can conceive of. In the case of the fast reactor it's possible to get a magnifying effect in the explosion by re-wetting an overheated core with sodium. I say it's possible because there isn't a sufficient amount of evidence positively to deny it. One can construct models even in relation to thermal reactors in which you can get energy releases which are broadly comparable. But it's more difficult to get a large nuclear transience inside a thermal reactor than a fast one.
CHAIR Would you feel sufficiently confident to build you fast reactors near centres of population?
FRANKLIN I think that as with thermal reactors it would be prudent to start off by constructing early prototypes in relatively remote sites and when one had got perhaps 20 years' experience to look at siting which was more nearly urban.
COTTRELL Is there an extensive series of pilot experiments being done jointly with the French and Germans on these nuclear explosions?
FRANKLIN Yes. The two main programmes which bear greatly upon a big accident concern the way in which gas is released from a fuel element when it is rapidly heated and the way in which a re-wetting of a core by sodium would or would not magnify the explosion.

In the course of his evidence (which mainly concerned nuclear proliferation), Professor Rotblat of S. Bartholomew's Hospital Medical College was asked about technological risks:
CHAIR Would you like to comment on the radiological

involvement?

ROTBLAT You have the Report from the Royal Commission which made some statements about present radiological hazards of nuclear energy in this country, and I fully concur. I myself am not happy about the present levels of radiation. I feel they ought to be lowered; but according to the present regulations I think that the Authority is doing their best to maintain the safety of their personnel. I do not subscribe to a statement made by the U.K.A.E.A. that the workers at Windscale have a lower than average rate of mortality, almost implying that a small amount of radiation is good for you; but neither do I accept the opposite report which came out a few days ago that in the United States similar workers have had an increase in the incidence of cancer. I think the samples are too small at this stage to enable us to express an opinion one way or the other.

The Friends of the Earth, instead of sending Submissions, allowed excerpts to be taken from their publications. The first is taken from *Energy Alternatives and Energy Policy. A U.K. Viewpoint* (1976) p. 5, para 14:

The fuel of a fast breeder reactor, unlike that of other designs, is mixed oxides of uranium and plutonium, and the core does not include a moderator. The failure modes of such a system are very complex. Spokesmen for the U.K.A.E.A. have recently gone on record as asserting that these failure modes do not include any which might lead to a significant yield of fission energy. In plain English they claim that 'a fast reactor cannot explode like a bomb'. However, in a memorandum dated 13th March 1975 Dr. Stephen Hanauer, one of the senior reactor safety experts of the U.S. Nuclear Regulatory Commission, drew attention to what he considered the unresolved problem of a possible 'core nuclear explosion' in a fast reactor. Clearly there is room for highly informed disagreement about the possibility. Needless to say the implicit risk is far beyond those associated with any other form of industrial installation. Assertions of confidence, however sincere, simply do not meet the case.

The second excerpt is from '*Non-nuclear Futures*' by A. Lovins and J. H. Price (F.O.E. 1975), p. 34:

The central problem of disposal, especially of actinides, is that the degree of permanence required subjects any terrestrial disposal scheme to geological requirements of which we have no experience and for which no responsible geologist can offer a guarantee: the time scale is less the realm of geology than of theology. Terrestrial disposal must therefore be "retrievable" in case of geological contingencies; and if it is "retrievable" to one sort of person it may be "retrievable" to others too, entailing "surveillance" on a time scale far exceeding the observed life span of human cultures. The wide array of technical disposal options proposed does not appear to meet this requirement of social longevity nor to take sufficiently into account its unique moral implications. So far, the only options actually available are for interim management until a safe method of disposal, if one exists, can be devised.

It appears that medium and low level waste management, like the decommissioning of nuclear facilities (few of which are designed to facilitate it), has received too little attention. The accumulation, over the next few decades, of tons of highly dilute and persistent hard alpha emitters (e.g. protactinium) seems to give particular grounds for concern, despite the U.S.A.E.C.'s long overdue 11th September 1974 proposal to eliminate the current practice of commercial soil burial of some transuranic wastes.

Mr Walter Patterson gave evidence on behalf of the Friends of the Earth:

SEARLE In the excerpt from the work of Lovins and Price they say that terrestrial disposal of waste must be retrievable, so that in the event of something going wrong people have to be able to get them back. Is this your view?

PATTERSON I haven't made up my mind. At the moment what we need is to guarantee retrievability for the next foreseeable future while we try and decide whether we can convince ourselves that we have devised a way which is fairly safe. At this stage I'm personally not convinced of this. There is an economic aspect here. If you could afford to spend an indefinitely large amount of money you might be able to devise a pretty impressively iron-clad way of isolating the material.

WHYTE What is meant by 'geological contingencies'? Are these earthquakes or what?

PATTERSON Not necessarily only earthquakes because we are talking about a very long geological time-scale. We're mainly concerned about the remobilisation of these materials into the biosphere so that they then become biologically active not necessarily in human beings specifically but in biological life systems. That could happen by incursions of ground water or by faulting, possibly even by an inadvertent man-made intervention at some future time.

ADAMS Are you aware of some prognostications by Dr. Bernard Cohen of the University of Pittsburgh who suggests that if the amount of waste from 400,000 megawatts was buried at a depth of 600 metres and started to leech straightaway, the resultant mortality in the community after one million years would be less than 2?

PATTERSON I've seen some of Dr. Cohen's papers. I don't find his line of argument very persuasive. I do know that efforts have been recently made in the U.S. and West Germany to devise accessible underground storage facilities for high level wastes and that the success rate has thus far been effectively zero because of the problem of unexpected holes which always seem to turn up in the salt domes which they have been investigating. But I'm much more concerned with the aspect of human fallibility than with the long term geological aspect.

McKNIGHT Do you think there's enough research going on into the geological movement of sources and flows of underground water?

PATTERSON No. A major I.A.E.A. conference in November 1975 offered a number of reports suggesting that these heavy elements migrate considerably and ground water is one of the most effective mechanisms of transport. This is certainly an area in which we need to know a lot more even in the U.K.

COTTRELL Were you aware of the investigations of the natural uranium reactor in Africa and the heavy elements from that hardly dispersed at all over a very long period of time?

PATTERSON There is one serious difficulty in using that as a parallel to a man-made storage facility. The material at Okra was geologically in place and was not disturbed both before and during the course of the nuclear reaction it underwent, whereas in order to set up the sort of storage facility we're talking about we must first intervene in the local geology if we excavate a salt dome or something of that sort, and this means that you then change the geostatic situation and you can no longer say that from then on the system is stable as it was before you intervened.

KEARTON If I produced a lump of plutonium on the desk like that, would you recommend that the hall was evacuated or would you think there was any hazard?

PATTERSON If it were nickel-plated which is customary or chromium-plated, I would be quite happy to hold it myself; but if it were not I would be a lot less happy.

RICHARDSON Do you think that the fast reactor in direct operation poses a particularly high safety problem or are you more concerned with the after-effects of a plutonium economy in the disposal of wastes?

PATTERSON Personally I am more concerned with the plutonium outside the reactor than I am with it inside the reactor because when it's outside it is accessible to human beings. What goes on inside a fast reactor are phenomena of very considerable complexity and I would not myself like to suggest that I have any kind of detailed knowledge of the reactor physics involved. I do know that there is a very pronounced and protracted high level disagreement among people who do specialise in this area as to the extent and kind of problems which might arise in the event of a malfunction in the core of a fast reactor, and I am prepared to wait

and watch and see how they themselves variously resolve this problem. There is an interesting overlap of safety with economics here. If you have one fast reactor and it begins to behave oddly you can safely shut it down without any hesitation; but if you are dependent on fast reactors for all of your energy supply or a significantly large part of it, and you find that you have a malfunction of this kind beginning to show up, you then have to decide whether you can safely shut the reactor down and cut off the supply of energy.
McKNIGHT What do you think would be the effect of erosion on glassified waste. Wouldn't it wear away?
PATTERSON I think it probably would. There is already a fairly large total quantity of active material dispersed dilutely throughout the oceans. The question you have to ask is whether the dilution and leeching would take place fast enough from locally deposited material to create a local concentration of radioactivity which would be dangerous during a short time interval. Over a long period I think you could confidently say that the total concentration of activity in the environment would get back down toward the natural level.
CHAIR (Question from the floor). In view of the known dangers to the lives of employees and to the environment of the coal industry is it not astonishing to find Friends of the Earth advocating a massive expansion of it?
PATTERSON We are not advocating a massive expansion but we are prepared to countenance the kind of expansion that the industry itself seems inclined to opt for. Our major concern should be to make better use of the coal that we do get out, however we get it out.

Professor Fremlin, Professor of Applied Radioactivity at Birmingham University was invited to make two Submissions, and allowed himself to be cross-examined:

(i) *Decontamination of outworn nuclear power stations*

If the very great capital cost of the CFR-1 has once been found and it operates successfully, its running cost will be so low compared with any other presently known or projected source of power that it is most unlikely to become redundant until some decades into the 21st century. If further breeder reactors are built they will last longer still, so the problem is not an immediate one. Nevertheless it is essential to have a satisfactory answer before CFR-1 is built.

When the useful life of a fast breeder reactor does come to an end, there will still be a great deal of valuable plutonium and uranium in the core, and in the surrounding uranium blanket, which would therefore be removed for processing with the high-burn-up fuel elements from fast reactors which are still operational.

We have therefore to deal with only radioactive nuclides in the supporting and containing structures. This will still at first be many millions of Curies but will be derived from far fewer elements and hence be simpler to deal with than the fission products from the actual fuel. The steel structures will be extremely active at first and are likely to contain large quantities of cobalt-60 derived from the minor cobalt component of the steel used and will contain considerable quantities of long-lived isotopes of nickel. These latter are not hazardous as they emit no gamma rays but the cobalt-60 does do so and, having a five-year half-life will persist for a long time.

None of the major elements in concrete will leave long-lived gamma-emitters but there will be small quantities of a great many impurities that will do so. Although it would be possible to disassemble the system quite soon, this would be a slow and expensive job, and it would certainly be useful to seal off the reactor vessel and leave it untouched for five or ten years until the major short-lived activities had gone. It would then be a purely economic question whether to keep it sealed off for a considerable further period, which would be neither dangerous nor inconvenient if newer reactors were being built on neighbouring sites, or whether to remove it completely and to decontaminate the site for other uses. The concrete sections could be buried untreated with glassified wastes after washing off any surface contamination and the steel sections and other materials could either be cast into waterproof concrete blocks and buried also or processed for useful materials before glassifying and burying the undesirable ones. Neither plan would cost more than a few tenths of 1% of the original capital cost of the plant. It may be noted that the steel and concrete of the main shielding container are likely to be considerably less radioactive than are those of a thermal reactor of similar power output, owing to the presence around the core of the thick uranium screen in which the breeding of fresh plutonium takes place.

In no case is there any reason to think that the demolition crews, working where necessary from shielded vehicles with remote-controlled tools, would be more at risk than are the crews demolishing any large building of an industrial chemical plant.

(ii) *Disposal of long-life toxic residues*

The construction and operation of fast breeder reactors will make very little difference to this problem. The thermal efficiency of CFR-1 will be about double that of the Magnox reactors and it will therefore produce about twice as much electricity for the same yield of fission products. As far as production of long-lived fission products are concerned, therefore, CFR-1 will be as good as the A.G.R.s and better than any other types of thermal reactor. It would even burn up a useful fraction of the long-lived isotopes that it produced.

Against these advantages is to be set the disadvantage that a fast breeder reactor would produce perhaps 20 times the total of alpha-emitting wastes in the form of heavy actinides produced by a Magnox reactor for the same electrical output. For the first few decades after removal from the reactor, fission products will still have greater activity than do the actinides, but after a couple of hundred years the longer-lived actinides will have the greater activity. They will still be active after 100,000 years although by this time their activity will be only twice that of the Magnox wastes. Actinides do not emit gamma-rays and are therefore dangerous only if ingested. Only one part in 10,000 is absorbed by the gut, so that they matter less than most fission products for a given activity level.

The actinides could be added to fresh fuel elements and returned to the breeder reactor, when they would contribute to the power output and themselves be destroyed, yielding much the same fission products as would be yielded by a similar quantity of plutonium. This looks like a good plan but we do not yet know how to do it efficiently. While uranium and plutonium are easy to separate from the bulk of the fission products, americium and curium, the important waste actinides, are not. It could be a long time before we learn how to separate them efficiently and I agree with the conclusion of the Flowers report that it will be better to consign them to safe long-term storage with the high-level long-lived fission product wastes rather than to keep them in the hope of employing them usefully.

Safe long-term storage is essential in any case for wastes from the existing thermal reactors and therefore neither the slightly smaller production of fission products nor the much larger production of heavy actinides is a serious factor to be considered in the decision whether or not to build CFR-1.

It has been suggested that the very high-burn-out fuel from CFR-1 might not be possible to process by plant now planned and that no satisfactory alternative is yet known. This is true but if, improbably, no solution can be found in the next fifteen years the spent fuel could be diluted with a few times its own weight of recycled uranium to give characteristics almost identical with those of spent fuel from the A.G.R.s.

SEARLE Will these long-lasting wastes have to be retrievable for a fair period in case the method of disposal is seen not to work?

FREMLIN I would myself choose to keep the wastes after treatment in tanks around Windscale available in stainless steel underground shelters as it were specifically for retrieval in case this should be desirable. I would also like to see permanent storage arranged.

KEARTON Am I right in thinking that you believe that the problem of toxic residues specifically related to CFR-1 has been exaggerated?

FREMLIN The Commercial Fast Reactor introduces a straightforward quantitative addition to the problem that we already have. It does not introduce a qualitatively new problem. That's all I'm saying.

CHAIR In your submission you say it might not be possible to process fuel from CFR-1 at the plant now planned and that no satisfactory alternative is yet known. What do you mean by this?

FREMLIN I mean that since the material has not yet become available, nobody has yet produced a final design on an adequate scale for processing it; and I was just guarding against the possible suggestion that we could not therefore run the CFR-1.

CHAIR You believe the design will be forthcoming?

FREMLIN Yes I do, but I don't want to have to say how quickly, and it isn't needed at once.

*Mr. F. R. Farmer, Safety Adviser of U.K.A.E.A. submitted the following evaluation of technological risks:*

Sir Edward Pochin has estimated the total risks in a nuclear power programme (N.E.A.(O.E.C.D. January 1976); mining and construction account for over 70% of the total risk of death or injury. The logical conclusion of this analysis is that nuclear power may well pose less risk than any alternative energy programme of similar size and capital content. This could be true, subject to the following:

1. That we have a well ordered society.
2. That the frequency and severity of accidents in nuclear installations roughly corresponds with the estimate of Professor Rasmussen—one which I believe can be achieved namely that the likelihood of an accident releasing 1% to 10% of the volatile fission products lies in the range of $10^{-5}$ to $10^{-6}$ per reactor per year. This is the current U.S. target for a large reactor programme.
3. That the risks in a Fast Breeder Reactor should not be greater than thermal reactors.
4. An adequately safe method of waste management be developed.

In discussing risks arising from accidents, it is not meaningful to say that an accident could kill or injure 1, 10 or 100 people. There is no easy way of defining a "likely" event or a maximum event, but rather a predictable range of increasing severity with decreasing probability. This applies to nuclear and to non-nuclear hazards, hence a framework is needed against which risk estimates may be presented and compared. One method is to present best estimates of hurt to the community, in death, injury and cost against the likelihood of the result. Techniques are in use and are being improved to do this:

> An estimate of the consequences of a severe accident to a reactor is given in the Report of the Royal Commission on Environmental Pollution—namely that 200-400 deaths could occur in the surrounding population within 10-20 years after the event. I estimate the probability of this to be less than 1 per million per year per reactor; a factor of 10 to 100 lower in probability than that a similar number of casualities might

arise from an accident in other industrial activities.

The surrounding population is also at risk from accidents at work, at play and on the road; within 10-20 years well over 1,000 will die from accidents (at present risk rates).

After a nuclear accident there could be contamination of the ground (mainly Cs 137); this would be easy to measure, would lead to great inconvenience, heavy cost in clean-up, but not to death or injury.

The development programme for the Fast Breeder Reactor has always aimed to show that it can be as safe as thermal reactors. The reactor is easy to control—it cannot behave like a bomb; the composition of the fuel guarantees a negative temperature coefficient. I cannot give a balanced safety review in five hundred words. I have presented many papers on the subject, the most recent at the U.K. Pugwash meeting (published in *Nuclear Reactors,* ed. J. Rotblat, London, 1977).

The usual practice in accident analysis is to make pessimistic assumptions in situations where we have inadequate knowledge and then direct research to eliminate or reduce those which have an important effect on the outcome of the accident; one such effect is the thermal interaction between overheated fuel and the sodium coolant. This is not a new problem. It is being studied in many countries and may be reasonably resolved within five years. If the effect is small, it would substantially increase my confidence in the safety of the Fast Breeder Reactor.

I do not believe the decontamination of nuclear plant to be a hazard—it is not a problem peculiar to a Fast Breeder Reactor programme. Power station sites may be used for one hundred years. The reactor is a small part of the site. At the end of life, a site would have all fuel removed leaving only irradiated steel and concrete inside a concrete box. The options are, to break up the box and contents and dispose elsewhere or to collapse the box on site and landscape. Either option costs money, the risks are small.

The problem of waste management is well presented in the Report of the Royal Commission on Environmental Pollution. The problems are not significantly different as between thermal and fast reactors. It is generally agreed that more work is needed to establish that negligible leakage can occur from vitrified waste cylinders.

As to the ultimate disposal to the ocean bed or on land—I hear views expressed by oceanographers and geologists about the uncertainty in any long-term estimate of behaviour patterns. I have no expertise in these fields but there will always be uncertainty in the analysis of the possible consequences of very unlikely events or of events many thousands of years ahead. There are major uncertainties about the effect on our universe of the survival of five to nine billion people for a few generations.

This leads to a wider consideration of risks to a community or the environment. Risks to people from hazardous industries are being studied in the U.K. A more thorough investigation of the effect of major schemes on the environment is now required in the U.S.A. under the National Environmental Policy Act 1969.

All such work requires effort; it is particularly important to achieve a sensible balance in the use of limited resources between nuclear and non-nuclear hazards and between those of higher probability and those of low probability.

Absolute safety is illusory.

Mr Farmer was cross-examined as follows:

RICHARDSON Is a fast reactor a more dangerous beast than a thermal reactor, or is it not?

FARMER It's very difficult to give a quick answer as the problems are different in fast reactors from thermal reactors. It has been said that a fast reactor can explode, but so can a thermal reactor; and the total energy availability in a thermal water reactor is higher than that currently assumed for accidents in a fast reactor, but this doesn't mean to say that one is equal to the other. The hazard to man from an accident in a reactor could be precisely the same from a thermal reactor or a fast reactor as it depends purely on the number of fission products that these reactors contain to a first order of approximation. So the difference depends upon the degree of credibility that is assumed in the accident analyses. We grew up with water reactors and gas cooled reactors from 1945 onwards, and it is only in recent times that we have considered at greater depth what we begin now to call the credible accidents. In water reactors this is the breaking of a pressure vessel. In fast reactors it's the nuclear coupled with metal-water interactions. But equally metal-water interactions are being considered in water reactors. I would say of fast reactors that we have spent more time considering remote accidents at this stage of its progress than we have of any other reactor; but we are uncovering things that we would like to know very much more about before we embarked on a programme of these reactors and these are specifically on its melt-down properties (that's the way in which the reactor core would disperse itself if it got very hot and the likely interaction of hot fuel and coolant). On both of these matters I believe that there will be a very sound technical basis on which we can come to some conclusion based on current research work both in England, France, America and Germany.

CHAIR (Question from the floor). The Royal Commission gave figures for a maximum credible accident following a release of only 10% of radioactive products from an F.B.R., although presumably an explosion could release considerably more than 10%. The figures given amounted to a possible one million casualties i.e. two orders of magnitude greater than the 10,000 given as the limit from an accident involving a thermal reactor. How could the result of such an accident be made acceptable?

FARMER The word casualties in the question from the floor is linked with the figure of 10,000 from the Royal Commission report which refers to possible cases of thyroid disorder not to deaths. I supplied these figures and I go along with them; but again I touch on the point of the degree of credibility which should be attached to these accident analyses. It is common to say that a water reactor or gas-cooled reactor will only release volatile fission products and these lead to a certain end-result. It is not absolutely impossible for it to release all its fission products. And we can devise accidents in which you could boil the fuel in a water reactor or gas cooled reactor. We regard these as being highly unlikely and we have more experience of these reactors. So at the present time we would say of a fast reactor if the accident were such as to disperse all the fission products that it would lead to a very very nasty accident and this would be equally true for a water reactor. The difference between 10% and 100% in my view is not so important. There comes a point at which a fairly large release of radioactivity becomes thermally self-uplifting by the generated heat of the fission products and a cloud so released will go up in the air instead of going along the ground and a release of about 1% to 10% is about the optimum to get the most number of casualties. Unless we were satisfied that these accidents which lead to a total dispersion of the core were so impossible by virtue of phenomena which we could then understand it is highly unlikely that we would ever proceed with these reactors.

RICHARDSON How safe would we be and how sensible would it be to go ahead in the short term when so much is unknown about CFR-1 in the long term?

FARMER I query the proposition that so much is unknown. We started work on metal coolant interaction 3 to 4 years ago and within the next 3 to 4 years will have a vast amount of information from work on Phoenix in France, and our own work. In the last 2 years we have started work at Harwell on the rapid pulse heating of reactor fuel heating (both in one second and in milliseconds) to find out the result. These experiments take some time to interpret but I believe in 5 years many of these questions will appear substantially different from what they do now.

*U.K.A.E.A. made the following submission on technological risks:*

Mortality among A.E.A. and B.N.F.L. workers is lower than would be expected from consideration of corresponding groups in the population as a whole. There is therefore no evidence that nuclear power is not a safe and healthy occupation. We have already 18 years' experience in operating fast reactors. Commercial fast reactors (most of the nuclear components of which are already working in our own reactors) will only be built and allowed to operate if they meet the safety standards imposed by the N.I.I. on all nuclear plant. The risks to personnel and to the community will therefore not be significantly different for fast reactors than for other forms of nuclear reactor.

2. Fast reactors consume as fuel both the plutonium which is created in the uranium fuel elements used in thermal reactors and also the uranium itself. This is the only way in which the bulk of the uranium fuel can usefully be used, effectively increasing the value of uranium fuel for nuclear reactors by a factor of about 50. (Thermal reactors can use recycled plutonium, but much less efficiently with only a small effect on uranium requirements). Radioactive waste from fast reactors is not a noticeably different problem from that arising from the reprocessing of thermal reactor fuel; similarly there need be no major differences between decontamination of fast reactor sites and thermal reactor sites.

3. At the present time when the irradiated fuel from thermal reactors is processed, the plutonium and depleted uranium is stored for use in fast reactors. The highly active liquid waste is stored in double-skinned tanks which are safe and secure. From the late 1980s, it is planned to vitrify these wastes and by the time the similar vitrified waste is produced in significant quantities from fast reactors, one or more of the disposal methods for solidified wastes, now being developed internationally will be available.

4. The probabilistic approach to safety matters has been set out in Mr. Farmer's note to this meeting.

Sir John Hill, Chairman of U.K.A.E.A. was cross-examined as follows:

SEARLE We're told that you can vitrify high-level wastes and then put them somewhere. As a geologist, it worries me where you're going to put them. They're safer because they're vitrified, but what about the rates of devitrification and remobilisation of the different sorts of wastes? In advancing vitrification as a good idea, how much does U.K.A.E.A. actually know about the rates of devitrification if you haven't got the equipment which you would need in order to discover what they are?

HILL We have done experimental work on this. Glasses have been irradiated at very much higher dose rates than the glass would suffer from the fission products. In other words we have been accelerating the damage processes to glass to see whether devitrification would take place and we're satisfied the glass will remain in an insoluble form for very extended periods of time.

CHAIR You haven't actually vitrified plutonium wastes yet, have you?

HILL Yes, we have. Vitrification was carried out at Harwell on a semi-technical scale. What we haven't yet done is to handle the full rate of throughput of the Windscale plant. So what is to be done over the next 8 years or so is to build full-scale plants to do what we know can be done on a small scale. The chemistry is extremely simple and could be done in any school laboratory. What is important is we get the engineering right to do it on a large scale.

KEARTON Would you regard the fast breeder reactor programme as being more efficient than a thermal reactor programme in the sense that it will in fact give less waste?

HILL Yes. The amount of waste that comes from a reactor is proportional to the heat produced and the fast reactors are more efficient convertors of that heat to electricity than the current generation of water reactors.

McKNIGHT In your normal civil programme (and of course in military programmes) there must be ample stocks of depleted uranium. Is it now possible and do you intend to re-enrich them with plutonium?

HILL We're at present storing the plutonium in very secure vaults at Windscale for future use in a fast reactor programme. There is I believe something like 20,000 tons of depleted uranium in drums and tanks in various parts of the country which if used in fast reactors would have an energy equivalent to 30,000 million tons of coal which is in excess of our coal reserves!

*British Nuclear Fuels Ltd sent the following Submission on technological risks:*

*History*

1. The nuclear industry had at its inception one great advantage over other well-established industries—it knew from the beginning that it was dealing with potentially hazardous materials and consequently the safety, both of the workers and the general population, has always been a prime consideration. Further, the implications, both physical and psychological, of the failure of a nuclear plant led to early adoption of a fundamental philosophy of making cost and convenience secondary to safety. As a result,

—the U.K. nuclear industry has had no fatal accidents due to nuclear causes.

—only 5 fatal accidents—all 'non-nuclear'—have occured in the U.K. nuclear fuel service industry, since its operations started 30 years ago. This is a remarkable statistic for any industry of its size.

—lost time for accidents and ill health of all kinds in B.N.F.L. is less than for any other comparable U.K. industry.

—B.N.F.L.'s health figures (for a workforce of some 10,000) are significantly better than the national average.

—no injuries have happened in the world wide transport of irradiated nuclear fuel.

2. The Royal Commission on Environmental Pollution commented on the exemplary record of the industry in the protection of the health of both its workers and the general public.

*Future*

3. The above record relates to a period during which health and safety standards have been progressively tightened. There is every expectation that future figures will maintain the reputation of the nuclear industry in these matters.

4. One other aspect of the nuclear industry worth quoting in the interests of perspective, is the saving of lives from using nuclear energy instead of other sources of energy. It has (for example) been calculated that the present United States nuclear capacity saves between 800 and 4000 American lives every year which would otherwise be lost, in accidents and from diseases associated with the use of fossil fuels.

*Radioactive waste management*

5. The problem of reprocessing nuclear fuel and storing nuclear waste are fully recognised but all facilities used are built to exacting standards independently approved and monitored by independent Health and Safety organisations. Present methods of storing waste have proved to be an entirely satisfactory and safe way of isolating it from the environment for many tens of years. The planned introduction of a vitrification process for highly active liquid wastes is a further illustration of 'design for safety'. Waste in virtually insoluble glass form could be stored indefinitely without needing the same level of surveillance. A major international effort is nevertheless being devoted to methods of ultimate disposal, which will provide permanent safe isolation. There is no reason to doubt that an acceptable solution will be found.

6. Much stress is placed by the critics of the industry on the need to store waste safely for 1000's of years (incidentally, the very long periods of radioactive decay only apply to a minute proportion of waste), but the toxicity of wastes from other industries (e.g. arsenic, lead and mercury) never decays. It tends to be ignored that the toxicity of plutonium in waste arising annually in the United States is less than that of arsenic in their pesticides alone.

Dr. D. G. Avery, Deputy Managing Director of B.N.F.L., was cross-examined as follows:

KEARTON In your submission you say that the lost time for accidents and ill health of all kinds in British Nuclear Fuels is less than for any other comparable U.K. industry. Comparable in what way?

AVERY With the total numbers of employees in the chemical industry generally.

ADAMS Currently there is plenty of space given in the newspapers to a leak that occurred at Windscale. Is the leak such that it could cause any accidents or be regarded as an accident in any Reports that you would submit?

AVERY In the light of what has appeared in the newspapers it will perhaps come as a surprise to people to realise that if all the water in the silo in which that fuel is stored escaped in one go and went off the site, that would be about equal to one day's permitted authorised discharge.

SEARLE Why then do you store it?

AVERY The reason we have water in the silo is that the wastes stored in it are old bits of Magnox fuel element cans, and if you leave Magnox fuel as a material dry it catches fire. It's nothing to do with radioactivity at all.

SEARLE But isn't the point at issue not so much the actual danger of the water that has escaped from the leak but the fact that there was a leak and that there may be leaks in the future? Could there not be leaks of a more serious character?

AVERY There are no absolutes. But the leak about which there has been so much publicity was from the first of these silos that we made. Subsequently their designs have been improved so that, for example, they have been given double walls and should the first wall leak then there are detecting instruments in the space between so that a leak would be detected even before it got to the second wall. This is symptomatic of what's been going on over the last 20 years; a steady progress in safety devices, with multiple containments, so that we are much more confident now than we had any right to be twenty years ago.

CHAIR (Question from the floor). If the contents of the leaking silo are similar in radio-activity to one day's authorised discharge, and if two workers became contaminated by a tiny part of it, then surely the authorised discharges should be questioned?

AVERY The authorised discharges relate to the radio-activity that goes out into the sea and is diluted by the sea. As for the contamination of the workers involved, in fact they got splashes, as I understand it, on their hands and their clothing, which was discovered on a routine examination and washed off by routine de-contamination processes. They washed, it is as simple as that. To argue that if all this material were collected in one place and—let's be figurative—if someone had a bath in it, scarcely gives a basis on which to establish the limits for discharge into a dilute ocean.

ADAMS Could you say what you are obliged to do in these circumstances?

AVERY We are obliged by statute to tell the Nuclear Installations Inspector, which we did. We are obliged by our internal agreements with our Trade Unions to tell them, which we did.

KEARTON In one of your comments you indicate that although plutonium is very unpleasant material, so also are a lot of other materials in common industrial use, and you mention arsenic, lead and mercury which never decay. Are you implying that the risk from plutonium is really only a small fraction of the risks already existing from other industrial sources?

AVERY I believe it is, yes.

SEARLE Do you take into account the presence of insoluble plutonium as a lung borne problem? Can one compare say the inhalation of lead with the inhalation of plutonium and say that lead's actually worse than plutonium?

AVERY In the physiological sense I'm not competent to answer, but the point I want to make is that the precautions taken to prevent people ever inhaling plutonium are such that the risk from plutonium is less than the risk from lead at this point in time.

CHAIR You say that stocks of plutonium are significant and that larger stocks will not present any fundamental change in the feasibility of adequate protection. But if we don't have a Fast Breeder Reactor presumably stocks will get much larger?

AVERY Yes, they will; and a great deal of plutonium would accumulate as a valuable energy resource and it would be held in store. Sir John Hill referred earlier to a vault. The size of the vault really doesn't affect in any significant way the measures one has to take to protect it.

RICHARDSON The Royal Commission paints in rather vivid detail what it would regard as the dangers to the world of a plutonium economy. It considers that the quantities envisaged would be so great as to change the whole order of discussion. There would be so much around that some would be bound sooner or later to go astray and this is a very dangerous unpleasant material. Do you think the Flowers Report has overpainted those dangers, or not?

AVERY Yes, I do.

*The following Submission was received from the Central Electricity Generating Board:*

The Board has full responsibility under law to operate its plant in a safe manner and, in its management of wastes, to comply strictly with authorised limits. Public safety has been, and will continue to be, dependent on the achievement and maintenance of the highest possible standards of design, construction and operation of fossil and nuclear power stations. The Board now has experience of over 150 reactor-years' operation in the nuclear field without any accident involving radioactivity having affected the safety of members of the public or the Board's employees.

Largely because of the responsible manner in which the Board's operations have been conducted and the effectiveness of consultations with Local Liaison Committees and other bodies, considerable public acceptance of nuclear power in this country has been built up. It is vital that it should be maintained. To extend this confidence to fast reactors it will be necessary to establish a satisfactory safety case for the reactor itself and the fuel cycle and to have satisfactory experience of prototype working. As a purchaser of the system, with unequivocal responsibility for its safety in operation, the Board would need to be assured about safety, reliability and cost of the fast reactor and fuel cycle before a programme of commercial reactors could be ordered with confidence. The objective of CFR-1 would be to help provide such assurances.

*The Nuclear Installations Inspectorate volunteered the following Submission:*

The Nuclear Installations Inspectorate is the body charged with controlling the safety measures of all the concerns which handle radioactive materials. There is general and specific legislation to back up its activities and it is part of the Health and Safety Executive (H.S.E.) which is vested with the powers and responsibilities provided by that legislation. The responsibility for authorising releases of liquid and solid wastes rests with the environmental departments such as the Department of the Environment and the Ministry of Agriculture, Fisheries and Food.

The N.I.I.'s function can be divided into two distinct areas: the first and the one of concern here, is the specification, enforcement and inspection of safety measures and controls for nuclear reactors, for the processing of nuclear fuels and for storage of radioactive wastes; the second is the inspection and enforcement of safety at all other work places where radioactive materials are present. This second function will not be dealt with any further here as it plays no part in the subject under discussion.

The main statutory basis for the first duties described is the Nuclear Installations Act of 1965. One of the keystones of this Act is the requirement for conditions to be attached to the Nuclear Site Licence that is necessary before any body (other than the United Kingdom Atomic Energy Authority or a Government Department which are required to work to equivalent standards) may build or operate a nuclear installation. The Act allows for such conditions as are considered appropriate in the interests of safety for any specific installations to be attached to the site licence to ensure the protection of operators and members of the public from ionising radiations.

More recent legislation—the Health and Safety at Work Act 1974—governing both the health and welfare of persons at work and the protection of the public from risks arising from work activities, necessitated amendments to the 1965 Nuclear Installations Act. The principal effect of these was to transfer the responsibilities for the safety aspects of licensing and inspection of nuclear installations to the H.S.E. from the Secretaries of State for Energy and Scotland. As part of the H.S.E., the N.I.I. now has a line of responsibility to the Health and Safety Commission, the members of which were appointed by the Secretary of

State for Employment. In this context the Inspectorate can enforce the provisions of the Health and Safety at Work Act on nuclear sites operated by the U.K. Atomic Energy Authority.

Site licenses are granted by the H.S.E. based on the detailed considerations and recommendations made by the N.I.I. To enable this to be done, an applicant for a licence must provide sufficient information to enable the Inspectorate to satisfy itself on the safety of the proposed plant and site. In this connection neither the H.S.E. nor N.I.I. are involved in political or other decisions leading to the type of plant to be selected. Their function is to ensure that any proposed plant conforms to the necessary standards of safety.

Once the licence has been granted the Inspectorate has full powers to maintain control over the design, construction and operation of the plant, and the licence conditions may be varied at the discretion of the H.S.E. The safety assessment of the design of the nuclear plant is backed up by inspection and during the actual construction and the subsequent commissioning and operation of the plant. Formal approval is also required for the station operating rules which specify the safe operating limits, the availability of essential safety equipment as well as the proposed testing procedures. Any proposed alterations to these rules during the operating life of the plant, must be approved by the Inspectorate.

Although the H.S.E. has direct responsibility to ensure that all necessary safety precautions are taken by the licensee both the Nuclear Installation Act and the Health and Safety at Work Act place the responsibility for protecting people from harm on the licensee. The licence procedure is not intended to dilute that responsibility; it is one of the means by which the H.S.E. carries out its regulatory duty. The very wide discretionary powers to the H.S.E. under the Nuclear Installations Act and the H. & S. at W. Act are available to ensure that any requirements laid down by the H.S.E. and the N.I.I. are met.

No representative of the N.I.I. was able to be present to speak to this unsolicited Submission. The Chairman commented as follows:

I'm sorry that in a matter of such importance the Nuclear Installations Inspectorate has been unable to speak to their unsolicited Paper. What inferences we draw from that I really do not know. But there it is. I think that you would like to know that Mr. Benn has addressed a whole series of questions to the N.I.I. and we understand that the answers will be given very shortly. There are some 19 questions which the Secretary of State formulated himself concerning (a) the comparative safety of thermal and fast reactors, (b) the kind of accidents possible with fast reactors, (c) the possible consequences of a fast reactor accident, (d) problems arising from the use of plutonium, (e) the problem of nuclear waste and (f) the purpose and acceptability of building a full size "demonstration" fast reactor. In addition to these questions there were some 23 questions sent in by Mr. Nigel Foreman M.P., and questions formulated by the Friends of the Earth (some of which we're discussing in these Hearings) were also included. It would have been good to know the views of the Inspectorate on questions of safety and technological feasibility, but we shall have to await the publication of their answers to questions submitted to them by the Secretary of State for Energy.[1]

**Social risks of CFR-1**

Consideration of social risks, as with technological risks, began by a consideration of the Sixth Report of the Royal Commission on Environmental Pollution. Three passages from that Report were brought to the attention of the panel:

(i) *General Evaluation of Social Risks*

A belief that the necessary vigilance and continuity could not be adequately guaranteed in any normal organisation led Alvin Weinberg to postulate a "nuclear priesthood"; this would be a dedicated, self-perpetuating body of people forming a technological elite which would be entrusted through the generations with the task of safeguarding society from the hazards of nuclear power. The idea of such a "priesthood" may seem untenable, but it is an indication of the extent of the anxiety felt by some responsible people about the hazards. Concern about security also led Weinberg to suggest that "nuclear parks" should be established. By this is meant the siting of reactors and related fuel fabrication and processing facilities in large, self-contained nuclear complexes in order to facilitate security arrangements, particularly by eliminating the need to transport plutonium.

Many people are concerned about the implications for society of the security arrangements that might become necessary in a plutonium economy. An effective security organisation could not be merely passive, simply reacting to events. It would need to have an active role (as was recommended for the U.S.A. in the Rosenbaum report); that is, to infiltrate potentially dangerous organisations, monitor the activities of nuclear employees and members of the public and, generally, carry out clandestine operations. It would also need to have powers of search and powers to clear whole areas in an emergency. Such operations might need to be conducted on a scale greatly exceeding what would otherwise be required on grounds of national security in democratic countries. The fear is expressed that adequate security against nuclear threats will be obtained only at the price of gradual but inexorable infringements of personal freedom.

We are sufficiently persuaded by the dangers of a plutonium economy that we regard this as a central issue in the debate over the future of nuclear power. We believe that we should not rely for something as basic as energy on a process that produces such hazardous substances as plutonium unless we are convinced that there is no reasonably certain economic alternative. In forming a view on the future need for large-scale dependence on plutonium, therefore, we should consider the possibility and implications of alternative options for energy supply. We should also be aware of the advantages that nuclear power offers though we do not think that these outweigh the risks we have described, they are real and should not be minimised.

(ii) *Theft of Plutonium*

The value of plutonium and its potential for nuclear threat would seem to provide ample motive for theft. Some people believe that as the quantity of plutonium in existence increases, together with the numbers of people concerned with its use and aware of its potentialities, a market for illicit materials is bound to develop, and theft to supply this market is inevitable. We emphasise again that we are considering a situation that could develop in the future. At the present time plutonium is stored in fairly small quantities at a few sites where appropriate security arrangements can be adopted. On the A.E.A.'s projection of nuclear growth in the U.K., however, the amount of plutonium produced by the year 2000 would be a total of 250 tonnes growing to more than ten times this amount by 2030.

The element would need to be transported on a considerable scale. The risk of its diversion would be particularly great if the element was in a form in which, given a knowledge of necessary precautions, it could be handled without danger. The plutonium in irradiated fuel can be extracted only by complex chemical reprocessing and the intense radioactivity is such that these operations could certainly not be undertaken illicitly.

The utility of the plutonium to a potential terrorist would be particularly great if it were shipped as a pure compound, oxide or nitrate. The latter form is especially hazardous from the point of view of dispersion.

Plutonium could be illicitly obtained either by the persistent theft of small quantities by personnel within reprocessing and fabrication plants or by terrorist action against these plants or during transport. The risk of theft within the plants is minimised by personnel checks and by strict security arrangements for access to the material. Elaborate accounting procedures are used to keep a check on the incoming and outgoing quantities involved in any process, and hence to detect any losses and to deter illicit removal. However, there are inevitable uncertainties in materials accounting especially at the initial stages of plutonium separation and as the amount of plutonium in circulation increases, quantities that would be significant for illicit purposes might be removed without detection.

The risk of loss through direct terrorist action depends on the detailed security arrangements adopted for the storage or transport of plutonium in relation to the organisation, expertise and ruthlessness of potential terrorist groups. Some people concerned with this matter have postulated groups consisting of about a dozen disciplined and well-armed terrorists, possibly assisted by "inside" information and support; the latter might be obtained by threat against personnel or their families. If such groups have indeed to be assumed then the security arrangements required in the plutonium economy would be extensive and their implications for society should be considered.

(iii) *Threats of Blackmail*

The construction of a nuclear bomb by a terrorist group would certainly present considerable difficulties and dangers to those attempting it. The equipment required would not be significantly more elaborate than that already used by criminal groups engaged in the illicit manufacture of heroin, but great care would need to be taken in the handling of dangerous materials and in avoiding accidental criticality. A substantial knowledge would be needed of the physical and chemical processes involved, of the properties of high explosives and of the principles of bomb construction. We have been impressed and disturbed by the extent to which information on all these topics is now available in open technical literature.

There seemed no reason to doubt that a sufficiently determined group with the necessary expertise could construct a very crude bomb which might explode with the force of a few tonnes of TNT. The amount of plutonium required could easily be carried by hand. Though extremely inefficient in nuclear terms such a device would still cause much damage and would create immediate radiation which would be lethal over a range of several hundred metres as well as dispersing radioactive material over a wide area. More doubt attaches to whether an illicit group could construct a weapon with a much greater yield, say 100 tonnes of TNT or more. There is some dispute about this possibility. From the discussions we have had we have formed the impression that the British authorities are less persuaded than those in the U.S.A. about the credibility of the construction of such a weapon. We felt it necessary to settle the matter in our own minds and we therefore consulted eminent physicists both in the U.K. and the U.S.A. who are expert in the subject. Their judgment was that the construction of a bomb that would give such a yield was indeed possible, though the actual yield would be very uncertain, for it would be as much a matter of luck as of good judgment.

We have concluded therefore that it is entirely credible that plutonium in the requisite amounts could be made into a crude but very effective weapon that would be transportable in a small vehicle. The threat to explode such a weapon unless certain conditions were met would constitute nuclear blackmail, and would present any government with an appalling dilemma. We are by no means convinced that the British government has realised the full implications of this issue.

Sir Richard Doll again gave evidence on behalf of the Royal Commission:

RICHARDSON We have been told that the Royal Commission has overpainted its description of a plutonium economy. Can you convince us that it hasn't?

DOLL I'm not an advocate, but we have had the report of the Fox Commission since our own Report, and that was able to provide in much more detail evidence which we hadn't been able to obtain and we think the Fox Commission has thoroughly borne out our anxieties. The Fox Commission was set up by the Australians to consider policy with regard to uranium mining and the impact that this would have on society. This Commission stated that the evidence points strongly to the conclusion that very destructive nuclear explosive devices can be made from reactor grade plutonium produced in power reactors operated normally. The physical data needed to make such devices with uncertain explosive yield but probably in the range of hundreds to thousands of tons equivalent of TNT (we'd given an upper limit of 100) are available in open literature. Construction of a nuclear explosive device is not of such complexity as to be beyond the apparent resources of existing terrorist organisations. Production of a device that dispersed plutonium oxide in the atmosphere would be comparatively simple. The evidence indicates that undetected theft of small quantities of plutonium from re-processing plants would probably be feasible expecially for trained organisations in countries with large nuclear industries. There is a very real risk that the opportunity and motive for nuclear blackmail will develop with time. This was an issue on which we were able to get very little help from our own authorities and most of the advice on this issue we had to get from individuals, some of them from the United States.

KEARTON Was it the terrorist risk which concerned the Royal Commission more than the technological risk?

DOLL Perfectly true. We didn't feel that readings of history justified us in believing that any country could be regarded as permanently politically stable over a period of more than 100 years or so in the future.

SEARLE The Royal Commission in their Report say they are by no means convinced that the British government realise the full implication of this issue which we've just been discussing and on which the Fox Commission commented after your Report. Do you think that the government has been badly or wrongly advised?

DOLL On this issue I don't know what advice the government have had, so I can't say whether it's good or bad.

SEARLE Did they really not play ball when you asked the same questions which you put to people abroad or to other governments?

DOLL We had difficulty in obtaining information on these very highly sensitive issues.

COTTRELL Is your worry about terrorists specific to the CFR-1 or is it a general worry about all types of nuclear reactors?

DOLL It's much greater with the development of fast breeder reactors because there's a much greater amount of plutonium being moved around the world and it's available in a form which could be used to produce explosive devices.

COTTRELL Does this mean that you're not convinced by the argument that fast reactors will be consuming plutonium whereas thermal reactors are creating plutonium?

DOLL No. That seemed a very naive argument to us.

KEARTON Are you saying that we should not investigate fully the CFR-1?

DOLL No, we're not saying that. We're quite satisfied that it should be fully investigated. What we are concerned about is committing ourselves to such a programme without satisfying ourselves that we have no reasonable alternative and that is what we think should be done in the interval during which the necessary experiments are being undertaken.

KEARTON Did you feel that there was no real attempt at the moment being made to find an alternative?

DOLL We did not think anything like sufficient funds or energy were being put into finding alternatives.

ADAMS Have you had complaints from anyone involved in finding these alternatives that they have been denied funds for experiments?

DOLL We had this at the beginning of our work, but by the time the Commission was ending fortunately many of these complaints had ceased. Mr. Chairman, may I add one other point? We haven't mentioned the dangers of nuclear war which in our view are greatly increased by the development of the uses of and the spread of plutonium throughout the world.

CHAIR Would you explain why they're increased?

DOLL Although we can be satisfied about the way plutonium is protected in our own country, we are not at all satisfied that it would be protected in the same way when it was being used in dozens of countries throughout the world, as it inevitably will be if the Commercial Fast Breeder is developed.

CHAIR Why should the development of the CFR-1 inevitably increase the proliferation of a nuclear economy overseas?

DOLL Once it had been demonstrated it can be done in this country it will be done elsewhere.

HENDERSON Am I right in thinking that a country which actually wished to make nuclear weapons and had a nuclear power programme would find it easier to make nuclear weapons if the nuclear power programme had extended to embrace Fast Breeder Reactors?

DOLL That is my understanding, certainly.

CHAIR I'm going to invite Mr. Avery to come back if he wishes. Would you like to comment on the suggestion that the Flowers Report has overstated the dangers of the plutonium economy?

AVERY The reasons why I believe the Report has overstated these dangers are threefold. The first is that there are unfortunately many other alternatives now open to the terrorist to wreak irreparable damage. I do not believe that the introduction of a plutonium system of fast reactors is going to represent in any way a unique opportunity for terrorist blackmail other than the encouragement which incessant publicity now gives to it. The second reason is that there are a number of measures that can be taken to make the material in transit and at any point where it is open to "attack" a great deal less attractive. But (and here I speak personally) I do not think it proper that these measures should be detailed in public since I believe that the gain that undoubtedly could result in terms of public confidence from open discussion of these details is like to be substantially outweighed simply by the ideas it would put into the heads of people who would not have them already.

If I might touch for a moment on the proliferation issue, the Fast Reactor, indeed a civil nuclear programme, is unfortunately not a prerequisite for nuclear proliferation. The technology is available now and a nation so minded could do it now whether we had a fast reactor or whether we didn't, whether we had a nuclear programme or whether we didn't. In so far as it has been suggested today that reprocessing fuel in this country might lead to proliferation abroad, let me make my company's attitude perfectly clear. We would not intend to export plutonium from this country, no matter what the source, to anybody who had not received the full endorsement of our government and of the international agency concerning .the motives they had for requiring the plutonium.

*Surveillance leading to possible totalitarian rule*

There are three main grounds for concern. The first is with the rights of employees who work in a plutonium factory. They will require to be screened before being employed, as they already are, and they may be subject to unusual surveillance during the course of their employment. Security screening is of course usual in some other areas of employment; though the need for it may be accepted by prospective employees, its more disagreeable aspects, especially the approaches made to friends, are resented by many. Security requirements are relevant to the question whether private, non-state, organisations and employees should be involved with handling of plutonium and other fissile materials in the nuclear fuel cycle.

The second issue is over the secret surveillance of members of the public and possibly of employees who may make "undesirable" contacts. The activities might include the use of informers, infiltrators, wiretapping, checking on bank accounts and the opening of mail, and they would be practised on members or suspected members of extremist or terrorist groups or agents of foreign powers who it was thought might plan an attack on, or theft from, a plutonium plant. We regard such activities as highly likely, and indeed inevitable. No doubt these methods are already applied to certain small groups that are regarded as dangerous, so that their use in relation to the plutonium threat would constitute nothing new in principle. It is a matter of degree, and the real question is the extent of the surveillance that might become necessary in the future if there were to be great reliance on plutonium. If there were a significant number of factions or individuals who might be prepared to use plutonium in threats against society, then widespread surveillance could scarcely be avoided. We find it hard to believe that such an intolerable situation could arise in this country, though it might do so in countries with repressive regimes. It must be remembered, however, that in considering the hazards of the plutonium economy we are concerned with conditions as they might be fifty or more years ahead. What is most to be feared is an insidious growth in surveillance in response to a growing threat as the amount of plutonium in existence, and familiarity with its properties, increases; and the possibility that a single serious incident in the future might bring a realisation of the need to increase security measures and surveillance to a degree that would be regarded as wholly unacceptable, but which could not then be avoided because of the extent of our dependence on plutonium for energy supplies. The unquantifiable effects of the security measures that might become necessary in the plutonium economy of the future should be a major consideration in any decision concerning a substantial increase in the nuclear power programme.

The third area of concern is over the action that would need to be taken to recover plutonium that was known to have been stolen, or in the circumstances in which a nuclear threat had been made. There would be a wide-scale and determined search for the missing plutonium, and general search warrants (which are at present illegal) might be required. The degree of disturbance to people in the course of a nation-wide search, or in the evacuation of a threatened area, could be considerable. There might well also need to be restrictions on the rights of movement and assembly and the suspension of habeas corpus if the threat of the plutonium being exploded were serious. Such powers already exist in theory under the Emergency Powers Act, 1920, but there might be doubts whether the occasion justified their use. It should be noted that the problems of enforced evacuation would also arise if there were a serious release of radioactivity as a result of a reactor accident.

*Professor Doll was not asked questions about this excerpt, but Mr. Grove-White, of the Council for the Protection of Rural England, was cross examined on his submission on a similar theme:*

1. Present security practices vis a vis the A.E.A.'s experimental fast reactor programme suggest that commercial dependence on such reactors—on the scale envisaged by both the A.E.A. and the Department of Energy—would necessitate rigorous security measures against theft of plutonium and sabotage of fast reactors. The seriousness with which such threats are already viewed may be judged from present safeguards, which include the use of security service vetting of A.E.A. employees, the arming of an independent A.E.A. constabulary to guard plutonium and its associated installations and extensive official secrecy on plutonium movements.

2. In *Nuclear Prospects,* we have suggested that a programme of 50 fast reactors (owned and operated by the Central Electricity Generating Board) could involve routine security service investigation of the political beliefs and personal associations of tens of thousands of non-government employees; extensive use of the Official Secrets Acts in the electricity supply industry; extended security service surveillance of political radicals and of conventional opponents of civil nuclear power (entailing phone tapping, informers and other techniques); and the need for an armed constabulary, thousands strong and outside conventional police controls, to protect nuclear materials in plant and in transit. Government—presumably advised by the security services—appear already to have decided (in the Atomic Energy Authority (Special Constables) Act 1976) that the effective operation of a nuclear constabulary necessitates operations outside the hitherto accepted framework for the bearing of arms in the U.K. In this sense, plutonium security may already be said to have had a serious constitutional impact in the U.K.

3. The seriousness of credible blackmail threats involving plutonium could create strong pressures for granting police powers of general search, to assist recovery of stolen materials.

4. The necessity for security against plutonium theft could entail increasing problems for parliamentary scrutiny of the continuing implications of a fast reactor programme. Executive secrecy might be easy to justify in the circumstances of such a programme.

5. The implications for civil and political liberties of an energy economy dependent on plutonium could give rise to quite new problems of public consent. Public concern could manifest itself at the time of electricity authority applications to build individual nuclear installations. Recent incidents at motorway public inquiries point in a mild way to the kinds of direct action that could arise.

KEARTON Are you against any security operations on the part of government of any kind at all? Is that a fair deduction?
GROVE-WHITE Indeed, it's not so. We thought when we wrote "Nuclear Prospects" that it was axiomatic that security precautions over Fast Reactors would be absolutely essential and we took the view that it was these necessary precautions which would be taken out of common prudence by the authorities to prevent any such incidents that would create social consequences which could prove themselves to be alarming.
McKNIGHT In examining questions of physical security, have you ever come across a case where somebody has got to carry out a purely routine drill and an incident resulted?
GROVE-WHITE There are incidents of that kind in the States. We have not come across such incidents in this country. One reason for this, possibly, is that it is very much harder to get public knowledge of incidents of this kind, particularly where they concern the fast reactor programme, because all employees engaged in the programme are obliged to sign the Official Secrets Act.
SEARLE We talk of the nuclear security force as a police force or as a military force. Am I right in thinking that both descriptions are wrong?
GROVE-WHITE Yes. The Atomic Energy Authority's Special Constables Act passed this year is an innovation in that it gives the right to bear arms to bodies other than the police and the army. The Special Constabulary is, in principle, a technical service of a nationalised industry and as such is subject to the same constraints about questioning in Parliament as any other technical service of the nationalised industries.
WHYTE Do you fear that in opting for the nuclear option, we may, without intending it, opt for an authoritarian rather than a free society in the long run?
GROVE-WHITE The aggregation of the particular measures that would have to be taken seems to point in that direction.

*The U.K.A.E.A. presented a Submission on the Social Risks of CFR-1:*

The Authority and other bodies in the U.K. have produced, handled and transported large quantities of plutonium, especially in physical and chemical forms specifically for weapons use, for upwards of 20 years without incident. Vastly larger quantities for a much longer period have been moved in the U.S. Such movements of plutonium have been made in bullion vans and under guard. Concern has been expressed that a large fast reactor programme would multiply such problems to an unacceptable extent. This is an inaccurate extrapolation of current practices leading to an incorrect conclusion.

It is common ground that a condition to be imposed upon a fast reactor programme should be that it did not have adverse effects upon the structure of our society. It is currently envisaged that a large fast reactor programme of perhaps 30 years hence would consist of perhaps two central reprocessing plants, similar to Windscale, servicing, say, 10-20 fast reactor power stations. The spent fuel from the power stations would be transported in the heavy shielded containers. No special security would be required since it is accepted that such irradiated fuel is of no attraction to terrorists or any organisation not possessing a reprocessing plant of similar capacity to Windscale. After reprocessing the nuclear fuel would be refabricated into new fuel elements and could then be activated to make the elements unapproachable, like spent fuel, and returned to the power stations in the same heavy containers in which they were brought to the reprocessing plant. If this were done only the fuel fabrication section of the central reprocessing plant would require to be a special security area and neither the fuel transport nor the fast reactor power stations would require any security measures other than those involved in the current nuclear programme.

2. *Material and Social Disadvantages in not Developing Fast Reactors*

If the U.K. alone took the course of abandoning fast reactors we should be cutting ourselves off from a world market for F.B.R.'s in which our share could have been very large because of the leading place that we now hold in fast reactor development. We should still not be insulated from terrorist threats, however unlikely these might be. But if F.B.R.'s were abandoned world-wide we would be increasing the strain on scarce and expensive supplies both of fossil fuels and of uranium. The effects would fall first and hardest upon the under-developed countries, whom we and other industrialised nations could outbid for what remained.

The associated waste of uranium (which would then be discarded 99% unburnt) would greatly outweigh all other world-wide efforts towards energy conservation with other fuels. Fast Reactors, by burning the uranium more efficiently, would raise the energy content of the world's currently known uranium reserves from the equivalent of its oil reserves to something greater than all the rest of its identified fuel reserves combined. They constitute the largest available conservation measure.

Sir John Hill again gave evidence for the Authority:

KEARTON Are you putting forward the fast breeder reactor as a possible benefit to under-developed countries in the future?

HILL I am not advocating that we build fast reactors in underdeveloped countries at this point in time. I am concerned that the wealthy countries of the world (particularly America but also Japan and Western Europe) have the money to buy and burn all the hydro-carbons in the world in two generations and I believe that we will find in 10 or 15 years' time that the poorer countries will not be able to buy the oil supplies that they require. I think that, if we take a very broad view about what would be for the good of the world, it would be that the countries which use most of the world's energy should conserve energy by using nuclear power and leave the oil and the gas for the other countries that are going to need them if they are ever going to dig themselves out of the appalling poverty that they are in at present.

SEARLE Are you implying that, if we were to develop Fast Breeder Reactors or increase thermal reactors, this would mean that we would not be in the market for fossil fuels, and so we would not be competing against less developed countries and would not outbid them?

HILL The Western World can only keep its economy going by consuming oil and gas at a vast rate and I think it would be better for the world if the Western World sets its economy in such a way that it didn't need to consume oil and gas at its present rate.

HENDERSON I wonder if the Atomic Energy Authority has done any calculations to suggest what might be the benefits which would accrue to the poorer countries of the world from this? Is it likely that these would be small in relation to other ways in which Western countries might act to the advantage of the poorer countries?

HILL We haven't done any specific calculations on this but I think it's clear that the Fast Reactor does provide the world with the ability to have perhaps 10 times more energy at its disposal than the entire fossil fuel reserves of the world.

COTTRELL Is the Fast Reactor a suitable energy source for making nitrogenous fertiliser which I believe is going to be the developing world's great need in the 21st Century?

HILL The Fast Reactor is going to be an extremely attractive source of heat and therefore will replace the use of fossil fuels for heat. The fossil fuels can then, in my view, better be used as feed to the fertiliser industry.

CHAIR (Question from the floor). Has the decision been taken to activate fuel elements so as to make them inapproachable?

HILL My view of the fast reactor programme of the year 2000 or 2020 (it doesn't matter very much) would be to have in Britain perhaps one or two factories like Windscale which would take in irradiated fuel from a series of fast reactor power stations. This fuel would be highly active, unapproachable by terrorists and therefore requiring no police and guards. At Windscale the fuel would be reprocessed and refabricated into new fuel elements. That particular operation, and that operation only, would be guarded by police. The fuel at the end of the processing line after inspection would then be irradiated for a short period in some small reactor located adjacent to the fabrication plant so that the fuel was too active to be touched and would then be returned to the power station in the identical 80-ton shielded container in which the fuel was brought to the reprocessing plant in the first case. If this were done, there would be no plutonium outside the Windscale gates which would be of any use to terrorists at all and I think this would avoid all the problems of a police state, with thousands of armed men, and we would, in fact, have to guard just one factory, Windscale, as we do at the present time.

WHYTE Is it not the case that proliferation of nuclear plant would greatly increase the number of people in this country who would be subject to security checks and the amount of activity that would be secret from people in this country?

HILL No, I don't think that is correct. The irradiated fuel that is transported around the country at the present time, principally from the nuclear power stations of the electricity boards up to the factory at Windscale, goes by rail and by road without any security guards. The reason for this is that the fuel is contained in 50 or 80-ton containers and these containers are absolutely secure so that even if a terrorist had by chance a 100-ton crane to take the lid off the container he would not be able to get access to the fuel because it is intensely radioactive. So it would be quite impossible for any terrorist or for any organisation to convert this fuel into a material that could be used for weapons purposes unless he had a clandestine factory of the complexity of Windscale. But this we regard as impossible and there is therefore no guard on this movement of fuel. It is true that we have security guards at our establishment at Windscale and at Dounreay because at both these establishments we are processing plutonium and there is a phase where we would not wish this plutonium to be stolen and therefore it is properly guarded.

*British Nuclear Fuels Ltd. included a Submission on the Social Risks of CFR-1, so far as the security of plutonium is concerned:*

*Stocks*

Recent debate on nuclear power would imply that the problem of handling plutonium was a new one, whereas plutonium has been produced since the first thermal reactor operated. Stocks of plutonium are significant now and larger stocks will not present any fundamental change in the feasibility of adequate protection.

*Terrorism*

The dangers of plutonium, rather than other more effective and easily obtained materials, being used by terrorists have been grossly exaggerated. Moreover, there are numerous ways in which the potential effectiveness of plutonium as a terrorist weapon can be reduced. It is terrorism rather than plutonium which we need to eliminate.

*General*

Irrespective of the timing of the 'energy gap', the recycling of uranium and plutonium in fast reactors has enormous economic advantage to the U.K. even if there are abundant alternative sources of energy. For instance, a single fast reactor with a planned power output of CFR-1, 1200 MW (E), would save the U.K. over £90 m. pa in foreign exchange on oil purchases as well as conserving natural uranium.

If the energy situation dictates that a fast reactor programme is necessary for the U.K. we believe there need be no serious misgivings about the consequences. Nothing can entirely be risk free but the responsible approach already demonstrated by the industry and the controls under which it operates will ensure that any risks are acceptably small. It has already been a major objective of the industry to conserve both the environment and exhaustible sources of energy.

Dr. Avery again presented himself for questioning:

KEARTON I've asked others, and now I'll ask you. The general trend of the evidence we are getting is that it is terrorism rather than plutonium we need to eliminate. Do you wholeheartedly agree with this?

AVERY Absolutely.

ADAMS Given the sort of progress that U.K.A.E.A. is suggesting for Fast Breeder Reactors, do you see the need for any massive increase in the number of armed police in the establishment?

AVERY A minor increase, perhaps, as the scale of operations increase, but certainly nothing massive.

ADAMS If we just stopped with the first Fast Breeder Reactor, what then?

AVERY No increase at all.

KEARTON The social risk of not going ahead with the CFR-1 would be the fact that we would have much more expensive energy as you see it?

AVERY Yes.

SEARLE Do you think that the addition to the police and the military of another group of armed people with specific security responsibilities is a good thing?

AVERY I think it is deplorable to need any increase in armed forces for this kind of purpose at all, but let us remember that we have accepted it, for example, at airports, and I do not find my personal liberties particularly infringed. I am much happier, I must confess, flying on aircraft after I've been searched.

*Earlier in the day, Professor Fremlin had answered questions on evidence he had supplied about the social risks of CRF-1: The Submission was as follows:*

The first risk I want to consider is that of theft of plutonium if the country comes to depend upon fast breeder reactors as its major source of electric power. The risks of theft of plutonium would not necessarily be increased by this but to avoid an increase the storage and transport of fuel elements might have to be under the same military control as is the storage and transport of atomic weapons. The building of a single fast breeder reactor however would not greatly change the existing situation. We have already considerable stocks of highly concentrated plutonium under civilian control, and adequate security arrangements to make attempts at theft both difficult and dangerous should have been easy to arrange.

The implication of large-scale military control of civilian power supplies are unpleasant, but it could well be valuable to see just what this means in practice with CFR-1. The necessary degree of military control is a vital part of the data we need before a decision can be made on the large-scale deployment of fast reactors.

The second type of risk arises from terrorist or military attacks on the reactor. The physical strength of the containing structures makes it quite impossible for one individual saboteur carrying a plastic bomb in a brief case to produce any danger to the public, although as in any major industrial unit he could cause very expensive injury to the plant.

It would, however, in my view be impossible to protect the plant with certainty from occupation by a sufficient force of terrorists prepared to lose a number of lives in the process. If sufficiently well informed they could then presumably engineer the equivalent of the maximum possible accident. I am not qualified to discuss this but it would certainly have to be sufficient to immolate the entire force of terrorists to be of much importance to the public at a distance, and the chance of failure would be great. I find it difficult to think of a credible origin for a large enough band of technically skilled and utterly self-sacrificing terrorists who would not prefer a far easier and equally lethal attack on a major chemical plant.

I cannot take seriously the idea of a war in which effective air attacks on a reactor could be mounted but in which we should not be attacked by genuine atomic weapons which would be put where there were a lot of people and which really would explode.

Professor Fremlin answered these questions:

CHAIR In speaking of possible civil disorders or war you say that terrorists could be expected to prefer a far easier and equally lethal method on a major chemical plant. Wouldn't you think that an assault on a breeder reactor would be credible in the case of Arab terrorists or even of some people in an island very close to where we are now?

FREMLIN I hadn't meant to say that it was impossible that

terrorists could be found who would be willing to immolate themselves in this way. What I mean to suggest was that if there were such bands of terrorists they could find much easier ways and much more spectacular ways of killing a great many people.

McKNIGHT You say that the physical security guard on materials can be strong enough to prevent terrorism in a civilised country. How do you deal with the case of a country to whom you are exporting reactor re-processing plants? Is there a case for an international police force to guard all nuclear plants?

FREMLIN I said that I thought the present security was probably pretty good but I did also say that there is no way of preventing a really adequate commando force of terrorists from beating our existing security, and this is certainly even more true of countries in which there is less experience and less certainty of controlling anything, let alone the security of these particular plants. The danger would exist if there were big breeder reactors in underdeveloped countries. But this doesn't seem likely to arise until after we have got a fast breeder reactor here, and that is the real question with which we're concerned now.

WHYTE I'm wondering what degree of military control you would find acceptable for the storage and transport of nuclear fuel, and at what point we might find ourselves with an unacceptable involvement of the military in civil life to the endangerment of civil liberties?

FREMLIN Yes, with a large number of fast breeder reactors this could be very serious, but I wouldn't be much concerned if one single Commercial Fast Reactor were built under military control. One would then gain from experience. It might do no more than show you that you shouldn't have a major breeder programme. That might be the answer.

WHYTE Is it possible to carry fuel for fast breeder reactors from one part of the country to another without military control with any degree of safety?

FREMLIN I don't know what they do with atomic bombs but they must carry those about . . .

WHYTE Are you suggesting that the state of the fuel for a fast breeder reactor is similar to that of an atomic bomb?

FREMLIN No, I'm suggesting that if we treated it that way, the thieves would steal a straight bomb rather than this much less dangerous fuel.

**Domestic Heating in the Future**

*Professor Fells, considered the situation of future domestic heating without CFR-1:*

Domestic heating is an area where established technologies can effect very considerable energy savings provided the financial incentive exists to implement them. By using heavy insulation and employing solar panels and heat pumps together with architectural design aimed at heat saving, the domestic heat load could be considerably reduced. In the event of energy rationing such techniques could substantially alleviate the plight of the deprived individual.

Professor Fells was questioned as follows:

HENDERSON We are enquiring about a specific area of energy consumption, domestic heating in this country. I imagine that the situation would be rather similar in countries which have much the same climatic conditions, and much the same standards of domestic heating. Would you agree that we ought to look at this sort of question from an international rather than from a strictly insular point of view?

FELLS Yes, I agree with that. If we don't build CFR-1 in 20 year's time, it is not going to make a profound difference, particularly if we stay with the kind of thermal nuclear programme that we've got. It is after 2000 that the real difficulties arrrive and the question which I was asking was this: what will happen, if for any reason we really seem to be running short of energy? One cause might be that we decide not to build a fast breeder reactor. Another might be that there will be a war in the Middle East in the next five years; and there are a whole lot of other possibilities. But whatever happens, if we start to run short of energy, people in their domestic environment will suffer. All I am saying is, they will probably suffer less than in many other areas provided they are prepared (or somebody is prepared) to spend some money. This is the chief difficulty at the moment. You can build a house or a building where you require almost no heat input at all, provided you construct it properly and provided you get architects who understand what the problems are and how to get round them. The technology is available to-day. Such a building has existed in Newcastle for 8 or 9 years and worked perfectly well as long as girls wore mini-skirts. The problems arose when they started to insulate themselves with maxi-skirts so that the heat input to the building was not as high as it had been before. Sixty-five houses with this low heat input or requiring such low heat inputs have recently been built by a colleague of mine in the Newcastle area. He tried to find central heating boilers with a low enough heat input to put into the houses and found not one was made in the United Kingdom. He had to go to France to buy it.

HENDERSON Could you say how much additional cost on average per house it would require?

FELLS I should have thought that to do the kind of thing that you might want to do, might cost anything between £1,500 and £2,000.

CHAIR Would the energy capital be repaid by the energy savings?

FELLS Yes, you would get a pay-back period of 8 or 9 years. Most people aren't prepared to accept a pay back of that period. Industry is rarely prepared to accept more than two years.

**Fuel and Agriculture in the Future**

*Professor Fells had expressed the following view:*

Fertilisers, weed killers and other agricultural aids are oil based although they could be made from coal. A physical shortage of these materials would seriously affect food production; alternative less intensive farming techniques are possible but could only be implemented on a long time scale.

On being questioned, he replied as follows:

COTTRELL You say that fertilisers could be made from coal. As I understand it, the amount of energy required to produce fertilisers for the third world in the 21st century will be something of the order of today's present world energy output. Do you feel that we can look to coal to provide energy to create fertilisers?

FELLS No, I suspect you can't look to coal to provide these things, and also to provide your electricity, although, of course, the amounts involved in fertiliser manufacturing are at present very much smaller.

COTTRELL Are we talking about world problems?

FELLS Yes. I am pessimistic about the world situation so far as fertilisers are concerned in 25 years' time, if we adopt a fuel economy which means that we've got to use our source of hydrocarbons for some other purpose, namely heating.

CHAIR Would you advocate, then, a massive increase in nuclear power to release a lot of these coal stocks for this purpose?

FELLS No, I don't advocate a massive increase in nuclear power. I am a sceptical optimist so far as nuclear power is concerned, and I prefer the kind of contained programme that I think I've been hearing about during the course of today. What I am really saying is that naive forward projections are going to break down for one reason or another, and perhaps we shan't be requiring—or probably we shan't go on having—this 5% annual growth, which we seem to anticipate, but that is a sociological judgement I think.

*Dr. Mellanby, formerly Director of Monkswood Experimental Station of the Nature Conservancy, submitted the following:*

Although primitive farming produced comparatively small crops, yields contained more energy (derived from the sun) than was expended by the farmer. Modern agriculture gives far greater yields, but it also uses much more energy, and some farming processes actually yield less energy than is put into them. Energy is used to drive tractors, to dry grain, to process crops for human and animal consumption, in intensive animal husbandry, to heat glass houses, to run factories building tractors and other machinery, and to make fertilisers and other chemicals on which modern farming depends.

A plentiful and cheap fuel supply would clearly reduce the cost of food production. A really severe energy shortage could make the highly productive form of agriculture found in Britain today difficult.

However, the use of energy by the farmer must not be exaggerated. Today all farming processes up to the "farm gate" use about three per cent of Britain's total energy budget. It seems reasonable to suggest that, even in times of severe energy shortage, this should be given priority, and that economies could be made in the heating of private and public buildings and in the unnecessary use of private and public motor vehicles.

Today the high yields, particularly of cereals, in Britain depend on the use of chemical fertilisers, particularly nitrogenous fertilisers. The production of such fertilisers uses nearly one million tons of fuel per annum. The yield increases of these arable crops are greatly in excess, in terms of energy, of that used in producing the fertiliser. In theory the energy wasted today when part of the wheat straw is burned is equivalent to several times that needed to produce all our chemical fertilisers.

Much of the agricultural research being done in Britain is aimed at reducing the use of energy, and there is no reason why the needs of farming should not fall substantially in the next 20 years. At present a good deal of energy is used for non-essential and luxury purposes, i.e. a third of the oil to heat glass houses. Also much more energy is used in processing food than in growing it. Saving from unnecessary packaging etc. could exceed the total "up to the farm gate" use with no serious harm to our diet.

Much of the energy used by farmers is to reduce manpower needs. If we increased our agricultural labour force (which many would welcome as a means of reducing unemployment) we would need less, not more energy use on our farms.

My general conclusion is that we could greatly increase our agricultural productivity and use substantially *less* energy than we use today.

Dr. Mellanby replied to questions as follows:

CHAIR Food is a great problem, it always is, but don't you feel that fertilisers and pesticides are going to be a great problem in the future?

MELLANBY No, I thought that the last speaker was talking nonsense on this particular point because we use so little energy to make fertiliser.

WHYTE You say that more energy is used in processing and in

packaging than is necessary. How far have political decisions, including, for example, the abolition of retail price maintenance, contributed to wasting energy in this regard?

MELLANBY We are using about 3% of our energy budget on all our food production up to the farm gate. We are using more than twice that between the farm gate and the shelf, that is to say in processing; and it is clear that at least half of that is counter-productive; unnecessary packaging, unnecessary sophistication, making the food less palatable and less digestible in many ways, but more attractive for selling. So if we are going to be short of energy, then we surely can get our priorities right and cut down on packaging. As for fertilisers, I think we are using in this country about the maximum amount we are ever likely to use. It is dangerous to foretell the future, but I would see a considerable decrease in the amount of chemical fertilisers and a greater use of organic manures many of which are now being wasted, so that the amount of energy used by agriculture is likely to go down considerably, together with a greater productivity than at present. But even if the whole of the world used nitrogenous fertilisers at the rate we are using them in Britain, it would be less than 100 million tons of fuel equivalent for the whole world. This is far less than is used by American motor cars and such like. In proportion to other uses, it is really very, very small.

COTTRELL If you go to the developing world, then will not the energy requirement for fertilisers and desalination of water and suchlike in order to increase the yield per acre there make a very heavy demand in the early 21st century?

MELLANBY Fertiliser will not. As for desalination, no one is going to use that for agriculture except under very unusual circumstances.

SEARLE If we increased our agricultural labour force, which many would welcome as a means of reducing unemployment, we would need less not more energy use on our farms. Sociologically as well as energetically, do you see this coming about anyway in the next 30 or 40 years?

MELLANBY I think that this is likely, and it would be desirable sociologically.

ADAMS In all sorts of industries, in factories or on the railways and elsewhere, we use energy to avoid having to do everything by hand. Do you find that deplorable?

MELLANBY No, it is a good thing that certain types of labour which are very tedious and very harsh should, if possible, be substituted by machinery. But there are many agricultural processes in which man could live a reasonably satisfying life, particularly when these are concerned with animals, which are at present being done with the use of much more energy derived from fossil fuels. Today, when we have 1½ million unemployed and at least 5 million in administrative posts doing counter-productive work, a lot of people might be doing useful work on the farm.

ADAMS Can you assure us that farm work isn't tedious or hard?

MELLANBY It's hard, it can be tedious; but I think many people find it very satisfying.

**Evaluation and Assessment of Risks:**

*Dr. John Habgood, Bishop of Durham, sent the following Submission, but he was unable to be present for questioning:*

It is doubtful whether a clear cut decision can be made on the basis of risk evaluation alone. Both sides can plausibly argue that hypothetical future risks, whether technical or social, are not unalterable quantities. Human beings adapt their technology and their social structures in ways appropriate to deal with risks which are actually seen to be pressing. The way in which the world has learnt to live with the bomb, though only marginally reassuring, demonstrates that safeguards can in practice be devised to cover a wide variety of risks, at least to the point at which public concern about them ceases to be generally expressed. Whether such complacency would survive a worldwide proliferation of nuclear capacity is another matter altogether.

There are, however other implications of the CFR decision, not to be classified as risks, which in my view have even greater ethical significance. Energy is not a fringe benefit but the resource on which the whole of an industrial society depends: the means of its supply must therefore have a profound effect on the character of that society.

A nuclear society would almost inevitably be expansionist. The prospect of abundant energy would provide a strong incentive for continuing to think of industrial and economic expansion into the foreseeable future. This, in turn, would entail an indefinite commitment to a highly advanced technology which could not easily be abandoned once it had been accepted, and which would need a broad base of technical competence to support it. A commitment to the CFR would thus be a commitment to a technically-based society like our own, only more so; and it would raise doubts about the ability of the other resources of the world to sustain it, especially if, as seems probable, many of the developing nations make the same choice.

A decision against the CFR would be unprecedented in that hitherto modern science-based societies have seemed unable to resist further technical advances. It would, however, allow much more serious exploration of the alternatives and is probably more practicable in Britain than in many other countries, in view of our enormous reserves of fossil fuel. It would not be irrevocable since it is almost certain that the CFR will be developed elsewhere, whatever our own decision, and the appropriate technology could, if the worst came to the worst, be bought at a later stage.

There is, at present, sufficient popular feeling about the CFR to make a decision against it politically possible. Such a decision could be made to have great long-term advantages if it were seen to signal a reversal of current trends, a recognition of a necessary element of restraint in our goals and a first move towards what the World Council of Churches have called "a sustainable society".

*The following excerpt is taken from a Friends of the Earth publication;* World Energy Strategies *by A. Lovins (1975), p. 113:*

The strong ethical ingredient of future-risk problems is perhaps most clearly seen in the perennial controversy, apparently as far from resolution as ever, over the rather arbitrary standards for human exposure to ionizing radiation. Critics of these standards argue, rightly, that safety has not been proved. Advocates reply, correctly but perhaps immorally, that harm has not been proved. Critics point to disquieting statistical trends—inconclusive but suggestive—and say that the burden of proof should rest on the innovator, as is theoretically the case with even the most beneficial drugs. Advocates can take comfort in the difficulty of demonstrating any real low-level effects, owing to the lack of controls in large populations and the lack of statistical significance of small long-term effects in small populations. And while several billion people are being asked to sign by proxy an irrevocable Faustian bargain whose small print they have not read and are told they cannot understand, those perhaps most affected—thousands of generations yet unborn—are not consulted at all. As Kneese stresses, this sort of ethical issue is beyond technical scope or competence, is too important to be left to experts, and requires instead to be resolved explicitly in the most open fora of society. This will be slow and hard, but is hardly a dispensable luxury. Low-probability but high-consequence risks, risks of unknowable size, and risks whose avoidance demands social stability into the indefinite future may be impossible to assess without social innovations and ethical conventions.

Mr. Patterson answered questions on behalf of The Friends of the Earth.

CHAIR How do you feel the risks should be assessed?

PATTERSON The strategy that I would opt for is one which goes for diversity rather than for narrowness, and I think that this is why I am uneasy about putting even into a single demonstration Fast Reactor the sort of resources and time and commitment which would be involved. To take that same amount of capital, that same amount of staffing, and the same amount of time particularly in the next 10-15 years and to spread it around into different areas of energy use and energy supply in connection with buildings and transport systems and industrial processes and to move on from the investigation of other forms of energy supply technology actually to the establishment of them—this seems to me to make far better sense if only as an approach by a 'safety in diversity'. If we get some parts of the package wrong we still have other parts to take up the slack, whereas if we put all our eggs into a fast breeder basket and the bottom falls out, we are in serious difficulties.

COTTRELL Wasn't it established this morning that without the reactor programme, by the end of this century this country from all its other energy sources would only have part of its present power output or energy output, so there won't be slack, there will be severe shortage?

PATTERSON I didn't get the impression of having established anything of the kind myself. I do feel that we have been aggregating energy far too much today. It is very important to recognise that there are different kinds of energy requirement for which we use uniquely different kinds of energy and it is not possible to swap nuclear electricity for petrol, certainly so far as road vehicles are concerned. There is no ready conversion from one to the other. We are going to have to diversify research and development into areas which do take account of this.

COTTRELL Our present energy consumption in this country today is 330 million tons of coal equivalent of which coal itself provides 130 million tons, well below half of that amount, and even if we got it up to 200 million we would still be 130 million short of the present levels, and by the end of this century North Sea Oil will be running low, so there is a great gap. There is not this slack that can be taken up.

PATTERSON I think that the coal and indeed the oil and natural gas which we presently use could be used much more effectively. That 330 million ton coal equivalent is primary energy requirement. We are talking about the way that energy is reaching its end-use; and it is clear to me, for example, that if we move in the direction of small total energy systems, then we will be able to get two or three times as much useful end-use energy from the same amount of coal.

ADAMS Could you tell me if anyone has, as you put it, put all our eggs in one fast breeder reactor programme basket?

PATTERSON The sort of sum which is being talked about simply for the construction of one commercial Fast Reactor is in the order of one billion pounds and that is just for the construction programme. I am bound to say that if we put that kind of money, which will have to come from public funds at this stage, into that programme, however much we hope we may accrue it over 30 subsequent years, we are not going to have that money available for other directions, and those other directions need support.

HENDERSON You say that you believe in the principle of

diversity but you follow this up by saying that you want totally to foreclose on the nuclear option. Is this consistent?

PATTERSON I didn't intend to suggest that I was foreclosing on options. The fast breeder as a concept still exists and will continue to exist. The existence of that concept in the forefront of our planning at the moment will foreclose most of our other options because we will pay little attention to them and give them very little effort.

*Mr. F. R. Farmer offered the following assessment on behalf of U.K.A.E.A.:*

Evaluate risk, compare, how to assess, ethics?

A simplified presentation—the numbers appearing are illustrative, not precise.

Risks in atomic energy and alternative energy sources, or industrial or social activities are varied, viz. death or injury in employment or to the public: loss of benefit: loss of amenity; some effect on the environment etc.

Death or injury may be random but predictable as for accidents in mining, construction, transport etc. at about 1 per 10,000 per year. There may be delayed effects as from radiation or exposure to some chemicals. Radiation effects are assumed to have a linear dose/effect relationship so that casualties arising from an accident will include numbers derived from populations carrying risks estimated as 1 in 10, 1 in 100 or 1 in 1,000 of some later disability or death.

There is seldom enough knowledge of low level exposure to toxic materials to carry out similar estimates even although some chemicals are carcinogenic and some mutagenic.

It should be noted that the average risk of accidental death from all causes is around 2 to 5 per 10,000 per year.

A more dramatic hazard is that of a serious accident affecting hundreds of thousands or "laying waste cities". Increasing attention is being paid to the study of accidents having very serious consequences even if of low probability of occurrence. In some non-nuclear situations, the assessment of probability/consequence may be based on accident history, viz:

The likelihood of an aircraft crash may be $10^{-6}$ to $10^{-7}$ per flight.

The likelihood of a random crash on any site depends on nearness to airfield or to air lanes etc.

The random crash on Hollywood stadium has been established to be $10^{-5}$ per year.

on a reactor site in the U.K. probably $10^{-6}$ to $10^{-7}$ per year or lower.

Other estimates may be made of the likelihood of dam failures, the chance of ship collision in the Channel, the chance of major fire in an oil refinery.

In the absence of direct experience of major accidents, we can try to predict the risk of their occurrence by analysing the various ways in which they could arise, such as the failure of electrical and mechanical components, failures in maintenance, a selection of which in various combinations could lead to a serious accident—generally in well-designed equipment, single failures should not lead to accidents.

Laboratory and field data for a wide range of equipment is accumulating, techniques for analysis, judgment in their use, are developing in several areas, chemical industry, nuclear, air transport—and will play an important part in reducing future risks in many other areas of high technology.

An accident to a nuclear installation, or to a tanker carrying toxic or inflammable chemicals, or to factories making or storing these, may have results ranging from zero to hundreds of deaths. Accidents which have occurred usually kill few people, the same type of accident under unfavourable conditions of weather, population distribution, topography etc., could kill hundreds or thousands—the risk is small requiring the combination of several factors each of a fairly low probability.

Comparisons of future risks can be made, the exercise may require considerable extrapolation and may give only an order of magnitude assessment. Such assessment should lead to a more rational judgment in areas where emotion runs high—it does not help judgment to say that a city may need to be evacuated and thousands may die without giving an assessed probability—this is a remote possibility but one which already exists through non-nuclear hazards.

The desire to avoid a low risk of a future catastrophe presents a dilemma. The reduction of risk requires effort and there are only limited resources available. Fewer people will die from unlikely accidents than will die from the frequent and random accident or the potentially avoidable disease or hunger. However there is a violent public reaction if many people are killed at the same time or place (Flixborough, Moorgate, Seveso).

The techniques used in assessing the consequences of an accident may provide a better appreciation of the threat of sabotage. Most attempts at sabotage of a nuclear plant will result in local damage but no serious release of radioactivity. There is a small chance of a trivial release of radioactivity causing inconvenience but little harm and a smaller chance of a serious release. It is wrong to equate sabotage with disaster—the effort to disrupt society could more effectively be applied elsewhere.

The assessment of harm from waste disposal and comparison with other environmental upsets is extremely difficult. I suggest some estimate can be made of the probability of dramatic changes in the earth's crust—some estimate can be made of the likelihood of meteorite impact on earth, the likelihood of a new ice age or of magnetic field reversal.

I doubt whether we know enough to predict the effect of chemical contaminants in air or water—I have no experience in these fields. I suggest, however, it is probably wrong to equate upheaval of nuclear waste—or their preferential leaching from buried waste—with catastrophe. Such events are likely to be identified, measured and, at some cost, adequately dealt with by future generations. That most accidents have generally minor consequences applies also—in my view—to nuclear waste.

I cannot comment on the remaining threat of nuclear war—this I believe to be the dominant factor in the present debate.

Mr Farmer replied to questions as follows:

CHAIR Do you think that risks can be quantified? Do you think that the subject is basically ethical or would you think that one is able to quantify risks in the kind of way that a decision can be made?

FARMER There are many parts of modern development in which quantification is helpful, particularly when comparing equal things, that is, the possibility of hurting people by moving in one direction or another. We have been searching for the last 10 years for a framework within which we can make reasonable statements. These techniques are developing. They are being used in the aircraft industry and in the nuclear industry and by the Ministry of Defence. If we can compare hurt either as immediate harm or later developed harm. But we can't strike a balance on equal things, that is, pleasures against harms.

SEARLE Would you agree with what the Fox Commission said in Australia?

FARMER I believe we can have entirely adequate, very safe nuclear installations and transport procedures. The issue still remaining is mainly that of sabotage or war. I am not so fearful of the sabotage threat to nuclear installations as might often be read in the Press. Most attempts to sabotage nuclear installations will merely shut the installation down because of the protective devices within it, and I agree with another speaker that saboteurs would be much better directed elsewhere.

CHAIR I am sorry that the Bishop of Durham isn't able to be here to speak to his submission. He concludes that it is very difficult indeed to decide on assessment of risks either way, but then he suggests that the main ethical questions lie rather in the implications of a decision about breeder reactors for the nature of our society, and that there is an ethical decision here about the kind of society that we want. Would you concur in that?

FARMER I haven't been able to read it yet, so I cannot comment. I do, however, believe that the lessons that we are learning from trying to develop nuclear power are extremely applicable to the development of modern societies with their technological complexities. The ability to estimate what might happen in the future or what might have a remote chance of happening in the future, is very important in building up our society. These techniques could well be extended beyond the immediate problems of engineering. In fact they are being developed into a study of failure performances of people in various tasks and I think that this form of extension is important in our developing technology.

HENDERSON Would you agree with the view that the threat of war might be appreciably increased by proliferation of fast breeder reactors in the world?

FARMER It would be very much a personal opinion. I believe that it is possible that there could be some increase with an extensive use of reactors and reactor materials in many countries; and I would not know how to balance this against the possible threat of nuclear war which might possibly result from a shortage of total energy in the world.

HENDERSON Should Fast Breeder Reactors, if they are developed, be confined to a small number of countries? Then there could be very strict controls on exporting and reprocessing.

FARMER I do not believe that the Fast Breeder Reactor will be developed in large numbers until we, collectively, are as certain of its safety as an installation as of any other reactor.

RICHARDSON People say that the improbable keeps happening and they point to amazing accidents like the recent DC10 air crash over Yugoslavia, the Brownsferry Nuclear incident, the Windscale fire and many others. They say that had you looked at them in advance you would say that they couldn't have happened; and yet those sort of things keep happening and therefore reassurances are not acceptable. How do you answer that?

FARMER The statement you have made requires very careful analysis. The fact that a DC10 crashed does not deny the present statistical rate of aircraft crash between $10^{-6}$ and $10^{-7}$ per aircraft flight. You can have very low statistical loss but still one will occur from time to time. That is consistent with statistics. The Brownsferry disaster wasn't a disaster but a fire. It was entirely predictable and had you asked me about the likelihood of a fire in a cable trench, I would have said $10^{-2}$ per year, and when you have got that number of reactors you expect to have a fire. And so the fact that a candle lit the fire was in my view entirely irrelevant. It was going to happen sometime and many of these so-called disasters or so-called accidents are predictable. But they did not develop to a disaster scale; and we, particularly in the nuclear industry and in the aircraft industry, are taking note of all the incidents which could potentially lead to disaster in order that our designs may be improved. Lord Hinton said in 1955 that the nuclear industry cannot live on the basis of disasters, although in bridge-building and dam-building more has been learned in disasters than from successes. We have had to move from that area in the nuclear industry to learn from the small mistakes that are made, from the pumps which aren't working to the switches that didn't work, so as to prevent disaster. This has been a whole new development of techniques.

*Dr. Ravetz, Secretary of the Council for Science and Society, offered the following Submission on 'A Comparison of Risks':*

1. *Introduction*

The simple comparison of numerical measures of risk as a special case of cost-benefit analysis, has long since been discredited. But risks can be compared, through an analysis of their *structures*. In this way we can clarify the issues for policy decisions, and exhibit the roles of science, politics and ethics.

For each case, I will consider the physical hazards, the social effects, the limits of assessment, and the means of protection.

2. *The Risks of the F.B.R.*

Here the physical hazard is the prime focus of concern. A nuclear reactor is a completely different sort of system from an organism, human or social. Its design has been *conceived* abstractly, and has not had a chance to *evolve* through long experience. It is *brittle* in response to disturbance, not homeostatic like an organism. Accidents not anticipated in the original design are liable not to be buffered, but to rip through the system to culminate in disaster.

The external effects of a B.R. disaster may be imagined as the "nuclear" case where Seveso was "conventional". This would be a major assault on our environment and social fabric, in some ways worse than conventional war.

In the assessment of technological risks, the calculation assumes the comprehension of the original safety design, and the effectiveness of human monitoring through the whole fuel cycle. All official estimates of probabilities of accidents and of their consequent damage are only *technological forecasts, not scientic facts.*

The threat of disaster from the installations is ever-present. Protection depends entirely on the skill of the original designers and the vigilance of safety staff. The record so far (as revealed at B.N.F.L. Windscale) is far from reassuring.

3. *Abstention from F.B.R.'s*

Here the physical hazard is not a threatened catastrophe but the possible absence of a particular minor source of energy. Since there is no assurance of the success of a Breeder programme, even this loss is not certain.

Human society has evolved for millennia usually in conditions of scarcity and frequently enduring social instability. Great civilisations of the past have indeed declined and fallen, but by failing to respond to a many-sided challenge, rather than by stumbling on one single obstacle.

Calculations of the social effect of a hypothetical future shortfall of this source of energy can indeed be made; but they may be ignored as totally speculative.

Our protection against the hazards of energy-shortage of whatever sort is in the social application of our essentially human ability of adaptation and innovation.

4. *Conclusion*

The issue for decision is which sort of risk, with which sort of protection we prefer. Any nuclear economy presents a constant and perpetual threat to the environment, kept at bay by a small group of designers and technicians. Abstention from the nuclear option involves a challenge, in which all our common technological and social skills will be engaged. In this way the choice is ethical rather than technical: where should we place our confidence.

Dr Ravetz responded to questions as follows:

SEARLE So far as protection from the threat of disaster is concerned, you say that the record so far revealed is that Windscale is 'far from reassuring.' Have you been reassured by what has been said here today?

RAVETZ Not entirely. I have been impressed by the studies that have been made by Dr. Wakstein in Leeds on the accident and incident record at B.N.F.L. His point as an engineer, is that there is recurrence of the same sort of accident time and time again. Corrosion incidents occurred several times, so did concrete failures.

CHAIR At the end of your submission you say that the choice is ethical rather than technical. What do you mean by that?

RAVETZ I mean that it is no longer acceptable to use the mathematical, logical model of the two sorts of risks in which you set it all up, all of the accident pathways with their associated probabilities and their associated costs, with two models, one on the left with fast breeders, the other on the right without them; and then you do calculations and come out with two numbers. Perhaps 10 years ago people thought that you could, in principle, do that sort of calculation on large-scale social questions but this is no longer credible. This is not to say that mathematical studies are irrelevant to the engineering problems of risk. They are essential; but if you take two global problems with a different structure, if you look at the structure of the two sorts of risks, you see that you cannot decide the matter on technical grounds. It is a question of which sort of risk society wants to accept.

CHAIR Can you explain what the two main ethical issues on either side are?

RAVETZ There is a technical basis to the choice. This is, in my opinion, that the breeder reactors in particular and nuclear reactors in general are systems, which are brittle under the disturbances that lead to accidents and they do not have a great deal of experience of accidents built into them. They are buffered from disaster only by the intelligence of the designers that first created them and emergent credible accidents. And secondly, they depend on the vigilance of safety staff to stop trouble occurring once it has begun, and this is why we will have a nuclear economy even at the reactor site which in some ways is the least problematic, knowing what we do about transit problems and waste disposal. Even at the reactor site we depend on the designers in the first place and the safety staff indefinitely.

CHAIR And on the other side?

RAVETZ On the other side we would have an additional strain on a social system which itself is undergoing all sorts of strains. This could be one factor among all the others. We must also decide whether we believe that there is enough intelligence and resilience and co-operation in society to cope with this additional possible strain along with the others.

WHYTE In your first option the threat to the environment is kept at bay by designers and specialists. But is this not true of any advanced technology?

RAVETZ Yes, but I think one can fairly say that a nuclear plant with its many dangers and varied types of toxicity presents this peril in a more accentuated way.

RICHARDSON Quite honestly I haven't quite grasped the point at present. Is it that because the nuclear power stations are brittle, they are somehow ethically undesirable?

RAVETZ I didn't wish to say in any way whatsoever that nuclear power is unethical. I am arguing here simply that when you finally make the choice between the two systems (and it is partly CFR-1, partly nuclear power in general), in the last resort you don't simply go to your engineers and ask them to tell you which is safer. It is not a problem that can be solved in a merely technical way.

---

[1]N.I.I. have subsequently published its replies in 'Some aspects of the safety of nuclear installations in Great Britain,' Health and Safety Executive, 1977.

# *Chapter IV*

# Feasibilities: Economic and Technological

After questions on the future needs and goals of society, and the risks involved in having, or not having CFR-1, it seemed logical to enquire next into the feasibility of the whole project. It was not easy to distinguish technological from economic feasibility, but so far as was possible the distinction was maintained.

*Mr Surrey and Dr Dombey of the Science policy Research Unit at Sussex University submitted the following:*

We are neither anti-economic growth nor anti-nuclear power. We see nuclear power fulfilling an important role in meeting long-term U.K. energy requirements, subject to a satisfactory solution being found to the problem of radioactive waste disposal. However, in several published articles we have expressed serious doubts about the wisdom of an early commitment to build the proposed CFR-1. These doubts are summarised briefly below.

1. *Energy Gap.* Five out of seven scenarios in the government report on Energy R. and D. were based on high energy growth projections. These five scenarios have been held to justify high priority for the F.B.R. Before these projections are accepted more work is needed to assess:

(a) Future energy requirements for the final sectors of demand, allowing for the effects of price increases, technical change, investment, and structural change in the economy.

(b) The possibilities for controlling the depletion of North Sea oil and gas so that essential U.K. requirements can be met until well into the next century.

(c) The prospects for electricity growth, bearing in mind that a "high-nuclear" strategy depends upon nuclear electricity becoming economic in a wide range of crude-heat uses where at present electricity is very uneconomic.

2. *Uranium Supplies.* Knowledge about the world uranium reserves recoverable at various price levels is very limited indeed. For many years demand and prices were low: there was no incentive to explore. Increased exploration, stimulated by rising prices, may yield new relatively low cost discoveries. In any case, large quantities are known to exist in low

concentrations in shales, granite and sea water; and these might be exploitable if uranium prices rise substantially. Before CFR-1 is approved much more needs to be known about the possibility that the U.K. might obtain a guaranteed long-term supply from Canada or Australia; and the possibility of developing thorium-burning reactors, possibly in association with Canada.

3. *Lead Times and Plutonium Availability.*Even if the "energy gap" projections were realistic, the rate of build-up of F.B.R.'s could not be as great as postulated in the U.K.A.E.A. Evidence to the Royal Commission on Environmental Pollution. Assuming that it would take 10 years to build and properly test CFR-1, and another seven years to build the first commercial F.B.R.'s, only about three F.B.R. power stations (about 3900 MW) could be in operation in the U.K. by the year 2000. With a small initial stock of plutonium, a rapid build-up of F.B.R.'s would require a large complementary programme of uranium-intensive thermal reactors to create the plutonium necessary for the F.B.R. programme. Alternatively it would require a much shorter plutonium "doubling time" than current plans allow for and whose technical feasibility is not yet proven. Therefore it is difficult to see how an F.B.R. programme could make the U.K. independent of uranium until well into the next century.

4. *Economics and Technology of the F.B.R.*

(a) The price of uranium ore represents a fraction of nuclear fuel cycle costs and only a very small proportion of the total costs of nuclear electricity. High fuel efficiency will therefore not necessarily mean the F.B.R.'s will be economic compared with thermal reactors: much depends on the comparative capital costs of the F.B.R. An American study[1] indicate that F.B.R.'s would be economic only if their capital costs were no more than 15-20% higher that those of light water reactors.

(b) There are fundamental differences between F.B.R.'s and thermal reactors. Although nuclear electricity was first generated in 1951 by a F.B.R., a commerceal F.B.R. would represent a new generation of reactor technology. The technology necessary to work at the very high temperatures and high neutron irradiation is still unproven, and the problems of working with liquid sodium as the coolant are also severe. Given time, these problems may be solved; but too little experience has yet been gained with the 250 MW prototype after months of technical difficulties to have confidence that the design could be scaled up five-fold without serious difficulty. Experience in building previous reactor systems and on many other big technology projects has been of major time and cost overruns.

(c) The advantages of a 1300 MW F.B.R. over a smaller one are unproven: they depend upon its operating reliability. If the 250 MW P.F.R. at Dounreay proves to be safe and reliable, that design could probably be stretched somewhat. This would avoid the long lead time involved in building CFR-1 and might well be the more cost-effective medium-term option if allowance is made for (i) the development costs involved with CFR-1, (ii) the risk that technological progress might make CFR-1 obsolescent by the time it comes into operation, and (iii) the increased system costs when a very large plant goes out of service.

5. *Research and Development*. The predominant share of every R & D funding in the U.K. is in nuclear power. Within the nuclear R & D budget there is already a dangerous concentration of the F.B.R., while other relatively uranium-efficient reactor options remain closed (e.g. High Temperature Gas-Cooled Reactors). While there is no automatic provision for government funds to be transferred from one heading to another, a decision to build CFR-1 would very probably mean that funds devoted to other energy technologies (including other reactor systems) would be substantially lower than if CFR-1 were postponed.

6. *F.B.R. Strategy*. With bigger resources and home markets the U.S.A., U.S.S.R., West Germany and perhaps France can more easily afford the large development expenditures, spread over a larger series of reactors, and descend the learning curve to lower costs. Even if the U.K. is held to possess a *scientific* lead, it is unlikely that the U.K. could maintain a lead in the *commercial* development of F.B.R.'s. Licensing from abroad might well be the best method of acquiring F.B.R. technology, rather than building CFR-1.

Mr Surrey and Dr Dombey were questioned as follows:

HENDERSON You refer to an American study that a F.B.R. is likely to be worth building only if the capital cost should be no more that 15-20% higher than for a thermal power station. What is the evidence for this and what evidence is there about, specifically, a British Commercial Fast Reactor? How much more than an ordinary nuclear thermal station do you think a C.F.R. is likely to cost?

SURREY It's impossible to answer directly, because there's no experience to go on. All we have experience of is the cost of building Advanced Gas-Cooled Reactors in this country and building Light Water Reactors in a number of other countries. There have been large cost over-runs on those. But I have no basis, I'm afraid, for saying how much more it would cost to build in this country a Fast Breeder Reactor, except that a report from West Germany has recently come to our attention, that the cost of building a 300 megawatt Fast Breeder Reactor has escalated to about three thousand dollars per kilowatt, which is over three times the cost of building a Light Water Reactor.

RICHARDSON In this American study to which you refer and which claims F.B.R.'s are economic only if their capital costs are 15-20% higher than Light Water Reactors, what are its assumptions about the future price of uranium? Is it based on a 'historic' ten dollars a pound or current thirty dollars a pound or the high cost of uranium we might expect if the world doesn't have breeder reactors?

SURREY It takes a range of costs between 8.5 dollars a pound of uranium, up to 15, and then on to 30, and up to 50 dollars per pound of uranium.

RICHARDSON Is it sensitive to the world value of uranium?

SURREY No, these are mining costs, and not prices that the uranium companies would ask.

RICHARDSON Is it taking in a credit for the value of the uranium fuel that's bred in the reactor?

SURREY No, it doesn't include either depleted uranium stocks or plutonium stocks or re-processing. It's purely a mining question on costs.

RICHARDSON If it doesn't take a credit for the value of uranium, is it at all relevant?

DOMBEY Yes, I think so, because the fuel costs are a minor part, compared with capital costs. Depleted uranium only has a value for F.B.R.'s where fuel costs are even less important.

ADAMS Despite cost over-runs on the Magnox, has it not produced electricity cheaper than any other principle source? Why should not a Fast Breeder still be economic in terms of electricity costs?

SURREY This is really a question of the difference between sunk costs and the cost of new investment. The Magnox Power Stations were put in many years ago and on historic cost depreciation they're bound to be producing electricity very cheaply. The question is whether one would go ahead now and build more Magnox Power Stations or build some other system.

DOMBEY As Professor Henderson is here on the Panel I should draw attention to his figures, that the A.G.R. programme has already sunk some sum around two thousand million pounds before any electricity has been generated by the A.G.R. The A.G.R. provides cheap electricity on recurrent costs. It doesn't provide cheap electricity on total costs.

CHAIR You refer to the difficulties inherent in unproven technologies. You talk of major time and cost over- runs. When I look at the submission of U.K.A.E.A. I see a rather different picture. There seems a divergence of evidence, and I'd like you to comment on it.

DOMBEY We wrote an article in the New Scientist on 10th December which goes into the technical problems which are the cause of the unproven technology for the Fast Breeder, which I can explain if necessary. Let me further refer you to a statement of a nuclear scientist who is as eminent a scientist in this field as one can find, Dr Edward Teller. On this issue he gave a talk at the American Embassy about two months ago, in which he said that more basic work on the behaviour of materials and associated topics is needed on Fast Breeder Reactors to understand the technology for building a new station. He said in passing that Fast Breeders were a good idea when they were first thought of thirty years ago, but they haven't fulfilled their potential. There may be a greater potential in the possibility of breeding in the thermal reactors, for example in the Canadian C.A.N.D.U. system or in the high temperature system; but there are little funds available, other than in Canada for this work. As for unproven technology in general, it's been pointed out that the fast neutron flux for fast breeder fuel assemblies will lead to materials problems not previously experienced and much more serious than those encountered in thermal reactors.

SEARLE You seem to be arguing that it would be sensible for the U.K. to back out of development of Fast Breeders in order to come in later, using technology developed abroad. Is this what you're really arguing?

SURREY We're not saying that we should pull out of fast reactor development. My own personal view—and I think perhaps my colleagues' here—is that we should carry on with R & D. Having got the P.F.R. we should wait and make sure that that is thoroughly tested, rather than go ahead now with CFR-1.

WHYTE How great are your doubts about the value of high energy growth projections, and how much would you wish to question them?

SURREY As an economist I question the whole philosophy of the energy gap approach which is to extrapolate past trends well into the future, and by the year 2000 one could finish up with a range of uncertainty of the order of two hundred million tons a year, on the basis of estimates which the professional forecasters might make. In other words, it reflects the range of forecasting uncertainty. In particular, the exercise was approached in such a way that it ignored the possibility of pricing at long-range scarcity values. It ignored the possibility of controlling the depletion of North Sea oil and gas, and it ignored the possibility of importing foreign oil.

CHAIR You say that, even if the U.K. is held to possess a scientific lead, it is unlikely that we could maintain a lead in the commercial development of F.B.R.'s. Could you justify that?

DOMBEY Advanced technology often is a sink of money rather than a source. Research and development costs, historically, have been enormous, and in projects of high technology—for example, in supersonic aircraft or in nuclear power stations—they've only been possible because they've been undertaken as part of non-commercial development.

*Mr M. Garner, formerly Under Secretary in the Department of Trade and Industry, submitted evidence on the possible licensing of future F.B.R.'s from abroad:*

Much of the debate on the Fast Breeder involves from about 1990 onwards of a world-wide energy gap—a gap between projected requirements and prospective supplies from conventional sources that can be met only by nuclear generation on a scale that, if there were general dependence on thermal reactors, would be incompatible with the foreseeable supplies of uranium; the gap has to be filled in large part, therefore, by fast breeders because they use much less uranium initially and are independent of it eventually.

With this assumption, the fast breeder is both indispensable and economic (because in conditions of shortage, uranium prices will be high) but without this assumption the fast breeder is neither necessary nor economic (because uranium prices will be low to moderate).

Prima facie, it may appear, therefore, that the case for construction of a fast breeder—CFR-1 in A.E.A. parlance—is determined by the judgement made on the likelihood of the energy gap, the possibility of which no prudent government can ignore. But this is not so: there is still the option of dependence on foreign technology—and there is still the requirement for systematic evaluation of the costs and benefits of this option as against the option of embarking on CFR-1.

Such an examination involves the choice of methods and assumptions and these are themselves proper subjects for debate. To show that an evaluation is feasible, I suggest a possible basis—namely, to compare the cost of embarking upon CFR-1 with the royalties that might be payable upon 33 GW of Fast Breeder capacity constructed in Britain to foreign designs by the year 2000 (the A.E.A.'s estimate of the construction necessary).

Evidence given to the Energy Resources Sub-Committee of the Select Committee on Science and Technology shows that CFR-1 would cost £600m, as against £400m for a thermal reactor—part of this difference is due to the fact that CFR-1 is the "first-off" station; the evidence also shows that, in terms of system capacity, the generating boards have no requirement for either reactor in the foreseeable future. If it is assumed that there would, nevertheless, be a requirement for the capacity in ten years' time, the cost to the country of CFR-1 (all risks ignored) is the difference between the present value of £600m. spent over the next five or six years and of £400m. spent over six to ten years from now.

I estimate this difference at £221m.

The A.E.A. itself has two current bases for royalty payments for the use of its nuclear technology: a 1967 agreement with the C.E.G.B. for the payment of the equivalent of 0.006 per kW of electricity supplied and a later agreement with the National Nuclear Corporation for 25 per cent of the net profits on the export of a reactor.

Doubling the royalty of 0.006 per kW (to allow for the fall in the value of money since 1967) and taking the A.E.A.'s forecasts of fast breeder capacity in Britain in 1990, 1995, and 2000, and relating them to the Department of Energy's forecast of the size of the energy gap in 1990 and 2000, I estimate the present value of the royalty payments for 25 years on the 33 GW of fast breeder capacity at £125m; but, using the export royalty and allowing for net profits of 15 per cent on an estimated supply price of £200 per kW for the nuclear component, I estimate the present value of the royalty on 33 GW of capacity at only £33m.

There is, of course, no reason why royalties eventually payable should correspond to either of these two current arrangements; and it is, of course, a matter of judgement to which arrangement, if either, the cost is more likely to approximate; but these arrangements do provide a possible indication of the range of magnitude of a royalty obligation; and there is the pretty positive indication that, by not building the CFR-1, at worst nothing will be lost and, at best, there will be a very substantial saving of resources.

RICHARDSON Can I ask about your conclusion, in which you say that by not building the CFR-1, at worse nothing will be lost?

GARNER One doesn't lose industrial capability by importing the technology. One can license the technology and still retain the actual manufacturing capability in this country. Secondly, the chances of being held to ransom seem at present to be very low because there are at least five countries that are actively engaged in developing fast breeder technology. So it's far more likely that there will be rivalry to 'dump' the technology rather than that the importer will be high-jacked.

COTTRELL Am I right in understanding that you would be quite happy if we bought the design from the French and built a French model instead of a home-made model?

GARNER If you mean, should we build a French model now, the answer is No, that would be ridiculous, in my submission. The case for building CFR-1 now is to acquire and prove the technology so that in the event of a requirement for a greatly enlarged power station construction programme around about 1990, the knowledge and capability would be available. The drift of my argument is that we can leave the other countries to get on, as they seem to be resolute in so doing, with proving the fast breeder technology, and if and when there is a positive requirement for building fast breeders in this country, or when we see that the need is certain, we can then set about arranging to import the technology.

CHAIR Part of your argument is based on the pricing of royalties by A.E.A. Do you know if these were priced on a realistic basis, or if they've been upgraded for inflation, or on what basis these were fixed? Do you feel that these royalty payments would be a realistic basis for considering royalty payments to other countries?

GARNER I'm tentative about these matters. What I am saying is that in this country at least there seems to be a confusion of logic. The case for the fast breeder, which is related to world energy requirements and world uranium supplies, is confused with the case for building a CFR-1 in this country now. To my mind they're quite separate. And as far as the figuring is concerned, I was concerned in my submission to give people some quantified basis of comparison between the cost of a decision to start on CFR-1 more or less right away, and the cost of an alternative strategy.

HENDERSON Wouldn't your argument apply equally to France or the United States, for example? Should we, so to speak, let them go ahead and waste money, or are you saying—by implication, though not directly in the paper—that possibly France or the United States or some other country may be in a somewhat better competitive position in relation to F.B.R.'s than this country?

GARNER I do not exclude the possibility that several other countries—France in particular, or even the U.S.A.—may be on the way to losing money with their fast breeder programmes. But I do think that Britain's non-competitiveness is the greatest, and that we are the ones who most certainly should withdraw. The main reason for this is that the other countries will have a much larger requirement for F.B.R. stations of the generating capacity that will have ultimately to be built if it ever proved to be necessary; that is to say, they will have a much larger home market, and the economics of these long-term, high capital cost projects are very sensitive to all the factors of the length of the development and the size of the ultimate market. We are a very small market; we are, in my opinion, a rather poor economy, and therefore our requirement for power stations is bound to be smaller than those of other countries.

KEARTON Why are you so confident that people possessing this technology will be so anxious—to use your words—to 'dump' it on receiving countries? Wouldn't they be in an enormously strong position, just as the O.P.E.C. countries have been, to make very severe demands indeed for their technology?

GARNER They might if one left the matter until the very last minute, until the 21st century; but at the moment we're in 1976, and it is plain that there must be great uncertainty about the extent to which there will, or will not be, an energy gap around about the year 2000. So what is involved is the question of the size of the insurance premium that one is prepared to pay. And the risk is not that the technology will not be available, but that one may be called upon to pay heavily for it in the year 2000. That is a risk which seems to me certainly not to be worth the extra cost involved in starting on CFR-1 now.
CHAIR (Question from the floor). Is not one argument for our own CFR-1 that it is likely to be a better and safer fast reactor than the French? If so, why ditch it now?
GARNER I can't take this insular attitude that other countries are indifferent to safety, and I think that there is just as much likelihood that countries such as the United States, Germany, France and Japan will arrive at as safe and viable a reactor as ourselves. I'm merely saying that our reactor is almost certain to be uneconomic.
KEARTON If there is going to be an energy gap in the 21st century, and the fast breeder reactor, if the claims for it are substantiated, would play the biggest single part in removing that energy gap, would it not be prudent for this country to be able to have an insurance against this possibility by proceeding with the CFR-1?
GARNER In all probability the technology will be available. I cannot see that our continuing with the technology is going to make the difference in the 21st century between the availability or non-availability of fast breeders.

*In place of a submission from the Friends of the Earth on the technological feasibility of CFR-1, an excerpt from their 'Energy Alternatives and Energy Policy (1976) was considered:*

Neither the technical nor the economic performance of fast reactors is likely to inspire confidence. The proposed development of an economic and social system involving transport of plutonium by the tonne inspires precisely the opposite. . . .

In the U.K. the Dounreay Fast Reactor went critical in 1959 and reached full power in 1963. The subsequent discovery of void formation in core components, leading to swelling and deformation, necessitated a lengthy outage. The Prototype Fast Reactor at Dounreay went critical on 3rd March 1974, some two years late, after problems with complex structures, particularly the reactor roof. More than two years later it was still plagued by problems, including the seemingly inevitable problems with steam generators. It had not thus reached more than 30 per cent of design power. *(By the time of the Hearings, full power had been achieved.)* The French Phoenix is frequently cited as the paradigm, free of blemish, which validates the fast reactor as an energy supply technology. But it is generally believed that the conversion ratio of Phoenix is at best about 1; and the layout of its cooling system, and other design details, will require fundamental redesign for a plant of commercial size. Furthermore, a report in *Nuclear Engineering International,* April/May 1976 (p.23) quotes the cumulative load factor of Phoenix to the end of December 1975 as 53.3 per cent—hardly impressive. . . .

The history of the world's fast reactors is predominantly gloomy, and punctuated with episodes of unwelcome drama. In the U.S., on 29th November 1955, the Mark II core of the Experimental Breeder Reactor-1 was destroyed by a fuel meltdown. The E.B.R.-2 went critical in 1963, but did not reach full power until 1967. The Detroit Edison Enrico Fermi reactor near Detroit, Michigan was originally intended as a prototype commercial fast breeder power station; but from its initial criticality in mid-1963 to its final shutdown in early 1972, it operated only sporadically, and bred only enough plutonium to make its slightly irradiated uranium blanket ultimately impossible to dispose of safely, either by burial or by reprocessing. It encountered problems of fuel swelling and distortion, problems of corrosion caused by sodium coolant, problems with fuel-handling gear and problems with steam generators. On 5th October 1966 a metal strip from a safety device came adrift in the core and blocked two fuel channels, causing a partial fuel meltdown which took more than two years to clean up. The reactor never came close to its design output; even its decommissioning has presented Detroit Edison with a lingering, intractable and costly dilemma. The Fast Flux Test Facility at Hanford has been under construction since 1970, during which time its cost has undergone an estimated 800 per cent increase. The Clinch River Breeder Reactor, planned for Oak Ridge, Tennessee, was originally estimated to cost $700 million; this estimated cost has increased in two years to over $1700 million, and the reactor is not yet even under construction. The consortium of participating electrical utilities, more than 300 in all, have not increased their collective contribution, which averages substantially less than $1 million per utility. . . .

Plutonium fabrication facilities have less than a reassuring record. The Kerr-McGee facility at Crescent, Oklahoma, which had been manufacturing fuel for the Fast Flux Test Facility at Hanford, was shut down in 1975, amid a bitter controversy over the death of an employee, Karen Silwood. She was killed in a car accident in November 1974, en route to meet an official of her union and an experienced reporter from the New York Times, apparently with a dossier revealing malpractice in the plutonium facility. No dossier was found in the wreckage of the car, and officials of the Oil, Chemical and Atomic Workers International declared their suspicion that her car had been forced off the road. In 1971 British Nuclear Fuels Ltd. entered a contractual agreement with the U.S. firm of Gulf United Nuclear Fuel Corporation, to carry out work on fuel design and plant design. On 21st December 1972 the Pawling, New York, plutonium fabrication facility of Gulf United sustained severe damage in a fire, in which two explosions occurred and an undetermined amount of plutonium was scattered inside and outside the plant. The plant was permanently shut down. So far as is known, B.N.F.L. have not commented on the experience of their U.S. associate.

Mr Patterson answered questions on behalf of the Friends of the Earth:

COTTRELL I didn't quite understand the drift of your argument because the whole point of putting up the Dounreay Fast Reactor was, in fact, to experiment, to find out what the difficulties were. But to say they found difficulties, and then use this as an argument against it, seems to me illogical.

PATTERSON No, I think the point of my argument was to demonstrate that not only in the U.K. but everywhere—including the Phoenix in France—the discovery has been that very serious technological difficulties have arisen, and they have arisen in ways which have not readily lent themselves to solution; the point being that if you do carry out an experimental design and find that the design gives rise to very serious technological difficulties, it may make you pause about whether you should proceed along that line of development, given the resources you have to commit to it.

COTTRELL Would you agree that the Dounreay Fast Reactor, having now got pretty close to its original design parameters, has demonstrated a satisfactory outcome of the scientific work that's been put in on it?

PATTERSON It's my understanding that the steam generators will have to be replaced, which suggest that we still haven't got that particular aspect of the problem satisfactorily corrected; and since, in the intervening period since this was written, the Phoenix has now had trouble with two intermediate heat exchangers, which are also similarly complex and difficult technological units, this suggests to me that we should at least hold off until we feel reasonably certain that these essential components can be constructed adequately.

COTTRELL Yes, I think that's accepted by everyone. But the components which have given trouble are non-nuclear.

PATTERSON Sodium water steam generators, whilst they are not strictly part of the nuclear system, are an inherent part of the liquid metal cooled fast breeder reactor.

RICHARDSON Bearing in mind that until very recently the fast reactor programme in the United States was given a very low priority, and that their technical approach has been different from ours, how relevant do you think that detailed U.S. experience is to the plans for what we're doing in Europe?

PATTERSON We have examined fast reactor development in every country where it is presently being pursued, and established that there have been major technological difficulties in each of these countries. So the mention of the U.S. difficulties here shouldn't be taken to exclude the difficulties that have arisen, for example, in the Soviet Union, or in France, or in West Germany.

SEARLE Now that the Prototype Fast Reactor has only just come up to its design power, do you think we should now go ahead with the Commercial Fast Breeder? If not, how much further experimentation in terms of time should we allow before we say yes, or no, or even consider the question?

PATTERSON It would not be unreasonable to wait and see whether the new steam generators to be built and installed and in operation by 1979 perform better that those which we have had in operation thus far, because this is a very critical part of the plant, and it has proved to be the most troublesome part of every single Fast Breeder of whatever national origin; and many design engineers and welding experts consider that the problem may prove to be completely intractable.

SEARKE So, the answer is, briefly, '79.

PATTERSON I would at least think '79, yes.

WHYTE On the one hand I hear the voice of those who have supreme confidence that the problems associated with Fast Breeder Reactors and other forms of nuclear energy can be overcome, and on the other hand there is the voice which you and others have expressed, that there are very serious problems indeed which ought to be a matter of concern. Is the difference here a merely scientific one, can you enlighten me about the basis for the two attitudes?

PATTERSON If I were convinced that this was the only direction in which it was worthwhile placing our efforts I would then be inclined perhaps more toward the direction of the nuclear advocates, because I would feel that it was all or nothing; this was the only thing we could conceivably do and we must therefore simply do it as best we can. But it is my opinion that we are considering here a resource allocation problem and a problem of effort allocation, and I can see many directions which seem likely to me to have a better opportunity of paying off for us than this one.

WHYTE Are you really saying that the difference is a matter of commitment?

PATTERSON Perhaps. As far as my own mental attitude is concerned I'm much more interested in advocating policies which seem to me to make the best sense.

KEARTON Do you really feel that if sufficient scientific resources were given to alternative energy possibilities, a success-

ful answer would be achieved?
PATTERSON Certainly.
KEARTON Aren't the views you've just expressed, matters of opinion, really?
PATTERSON I don't think they're matters of opinion any more than some views about the feasibility of the nuclear direction are matters of opinion. They have been backed by economic and technological investigation of some considerable detail, not necessarily in the U.K., but elsewhere, and some early demonstration of quite convincing feasibility on an engineering level, and the economics look to me equally persuasive.

The following statement was received from the Central Electricity Generating Board;

The cautious approach suggested by the Royal Commission on Environmental Pollution is accepted. Decisions on fuel and energy, where uncertainty about the future is an outstanding characteristic, ought to be taken step by step after careful appraisal of all the available options from the point of view of economics, long term fuel availability, safety and the environment. Flexibility should be the key-note in future energy strategy.

The board recognises the difficulties involved. While the basic concept of the fast reactor is feasible, its translation into commercial plant presents many problems of technology and safety both for the reactor itself and the fuel cycle. The time scale for development of a commercial product that will conform to U.K. standards on safety, reliability and flexibility of siting is bound to be lengthy. A series of demonstration plants will no doubt be necessary before a full commercial programme can be adopted. Such development is not likely to take until the 1990s, and in the intervening period it will be necessary to continue to build thermal reactors. Because of the lengthy lead times involved a decision by the Government on the policy to be adopted, including the possible construction of a full scale demonstration reactor, should not be long delayed.

The costs involved in research and development are bound to be high and the Board believes there is a strong case for sharing the costs with overseas partners on an international basis, quite apart from the need to harmonise safety work and standards internationally. Other countries are now proceeding with demonstration fast reactor projects and delay in reaching a decision in this country could jeopardise the opportunity of making satisfactory arrangements for international collaboration. The United Kingdom has pioneered the development of fast reactor technology and is therefore in a relatively good position to meet both its domestic needs and to export reactors, fuel and components.

The U.K.A.E.A. offered the following submission of feasibilities:

Before commenting on the technical and other feasibility of programmes of nuclear power development, it must be made clear that the Authority do not propose programmes of power station installation. In the context of Fast Breeder Reactors, the Authority's proposals are that work should proceed on the design and development of a single full size fast reactor—which has become known as CFR-1—to act as a "lead" or "demonstration" reactor; and that the U.K. Government should state its intention—subject to satisfactory progress with design and development leading to a case for its construction in due course—to build such a reactor. This would enable the U.K. to play its full part in programmes of international collaboration of Fast Reactors. The Authority have not envisaged any commitment to the large scale use of Fast Breeders at this time. CFR-1 is a necessary stage to reach the objective that safe, economic, and reliable Fast Reactors will be available to us when needed. Given continuing and expanded exchanges of technology and experience, this objective is well within reach.

2. In considering the practicability of having a number of Fast Breeder Reactors on-line by 2000 A.D. and a large number by, say, 2010, the following factors are relevant:—

(a) Fast Breeder development has been proceeding for 25 years. Its eventual need was foreseen by the pioneers of atomic energy. The experimental Dounreay Fast Reactor (15 MW(E)) has been generating power since 1962; the Prototype Fast Reactor (250MW(E)) this year. There is a great deal of experience of handling sodium, of materials performance etc.

(b) The Prototype Fast Reactor was designed with scale-up in mind. Full size fuel assemblies are used; and the evaporators on CFR-1 will be the same size as P.F.R. Scale-up on other components is not large, e.g. a factor of 2 on pumps.

(c) For more than 15 years there has been a co-ordinated and integrated programme of research and development involving British industry, Harwell, Dounreay, and several other atomic energy laboratories.

(d) There will be a single, evolutionary U.K. design.

(e) Other countries are building the same basic type of F.B.R.

(f) There are no basically new component problems, such as large pressure vessels, although a lot of know-how is involved. British companies have built P.F.R.

(g) The ratings of CFR-1 are no higher than those of P.F.R.: the operating temperatures will be lower.

3. The capacity of U.K. industry has to be considered, remembering that there will be international trading of components. Demand might require two fast reactors per annum by 2000 (2.6 GW(E)) plus a similar amount of thermal reactors. 5000 GW per annum—a rate of ordering of

power stations briefly exceeded in the 1960s—should be within the industry's capability at that time:

(a) provided that ordering rates in the next few years sustain a viable power engineering industry in the interim
(b) provided there is a minimum number of designs
(c) remembering that the unit size of reactor and generator is likely to be at least 1320 MW and not many units will be required.

4. There is sometimes misunderstanding about the possibility of financial and economic constraints on nuclear power development. If nuclear power is preferred, it will be because its total resource costs over some reasonable period are lower than those of the alternatives. They are lower because the extra capital costs are more than offset by avoiding the alternative heavy fossil fuel costs which would represent a greater burden upon the economy.

5. The total cost of electricity production in 1975 (capital charges and running costs) was about £2000m—1.85% of G.N.P. If—just for illustration—we take an electricity system in 2000 A.D. roughly double the present size and assume that half the plant will be nuclear, then—taking current costs of nuclear and best fossil-fired power—the total cost of electricity production would be £3500m. An average G.N.P. growth rate of only 2¼% would be sufficient to keep this total cost at 1.85% of G.N.P.

6. Even on a straight comparison between capital costs, nuclear power presents no special problems. When total capital costs are considered—for fuel production and electricity distribution as well as generation—the capital costs of a wholly coal-fired and wholly nuclear-fired system are likely to be within 15% of one another.

Mr Blumfield, Director of Dounreay Experimental Station, answered questions on behalf of U.K.A.E.A.:

CHAIR You will have heard Dr Dombey and Mr Surrey giving their evidence, and you must have noted a disparity between their views and yours about the difference between the P.F.R. and the C.F.R. Would you like to comment?

BLUMFIELD The P.F.R. was built for £44 million and the cost therefore is £176 per kilowatt. That includes the price of a very large amount of experimental equipment as well as the generating side of that reactor. So, effectively, this country has got a very cheap reactor.

CHAIR Does this include insurance costs? Does this include research and development costs?

BLUMFIELD No. It includes the cost of *extra* R & D equipment. It also includes the cost of 100 miles of transmission lines from Dounreay.

SEARLE Would you like to comment on whether we should proceed now with the design and construction of CFR-1?

BLUMFIELD Firstly, the design of CFR-1 is well along the road, based on experience. That boiler experience has been very important to this country. We are the only people in the world that have faced this issue and we do know the answers to the problems, and that is the reason why the design of the boilers has been changed for the CFR-1, to avoid the problems. So, effectively, the CFR-1 design is well along the road because the P.F.R. was designed with that commercial fast reactor in mind. The fuel elements are exactly the same size; there is no extrapolation in size required for the Commercial Fast Reactor. There's no extrapolation required for the boilers. The only extrapolation that's required is on the pumps and the size of the vessel in which the reactor sits. And those are not really advanced technological problems.

SEARLE So you say we can go ahead now?

BLUMFIELD Effectively, yes.

COTTRELL What I found missing from your paper was a programme of dates.

BLUMFIELD Well, we are in the year 1976. If we had a decision shortly to go ahead, I think it would be about three years before the site work commenced. It would take us about ten years in the industry at the present moment to produce a system which will be operational. So the demonstration Commercial Fast Reactor would not be in full operation until something like ten to thirteen years from now. That is the situation. Now we have not, in the Atomic Energy Authority, determined a model for the future of development into the true commercial stage of the fast reactor. We have said we wish to take the demonstration model to an operational stage to show that it is economically viable, and to disprove some of the comments that we've heard today.

COTTRELL How long would be needed between the switch-on day and the day when you can say to Government, 'Yes, it's all right'?

BLUMFIELD My own personal view is that you'd probably want to run the Demonstration Reactor for something like three to five years to be certain of this. The demonstration would take between ten and thirteen years, and about another five before the first truly commercial station.

ADAMS From your experience at Dounreay would you say that some F.B.R. problems are incapable of solution and, if not, would you explain what problems you've had which you believe have been capable of solution?

BLUMFIELD It was predicted that the steam generators were one of the real technological problems with Fast Reactors. The design was carried through on a basis of experiments that were done in laboratories. The final boilers that were received at the P.F.R. at Dounreay did have some very small leaks.There have only been ten leaks in ten thousand welds, and those leaks are smaller in diameter than a hair. Let's get the issue quite clear. There have been no accidents from them; those leaks have been found by the leak detection system; the things have been shut down and dealt with. Now we've done a lot of R & D in the meantime to find out precisely why. R & D has shown us that we should change the design to make this a better one for CFR.

HENDERSON In the case of the Breeder Reactor particularly, we've had very little evidence to show that the resource costs are, in fact, likely to be lower than alternatives. When can we reach the stage when you and possibly the Generating Board as well can give some estimates of likely capital costs , of the sensitivity of the programme to possible changes in uranium prices, and also to future operating performance and the technical problems which may bear on this?

BLUMFIELD Certainly there are a number of figures that are floating around; I mean, the Demonstration Commercial Fast Reactor is being quoted at about six hundred million pounds for a 1,300 megawatt station at January 1975 costs. Before we finally get realistic figures on all this we have to take into account that the Nuclear Power Company has to finish off its design, and the C.E.G.B. have to be involved in this. At that point I am certain that you will be quoted figures which are correct.

RICHARDSON Professor Whyte raised the question with the last speaker that I'm sure is puzzling a lot of people; it's certainly puzzling me. On the one hand we have speakers like yourself who speak with absolute confidence about the fact that the problems are solved or soluble; and on the other hand we have speakers like the last one, Walter Patterson, who are very equally clear that there are major unresolved problems that we do not know yet how to overcome. While we hear conflicts—the two sides might almost be talking different languages—the public and the Government are almost bound to conclude that the truth must lie somewhere between you. If that is so, we end up half-sceptical about nuclear power, and if we are half-sceptical, we are not going to build it. It seems to me, therefore, to establish your case you've somehow to show—beyond a shadow of doubt—that the sceptics are wrong and that you are right. How on earth are you going to do that?

BLUMFIELD Well, I've been trying for something like five years by writing to the Press, by having interviews with the B.B.C., trying to get my points of view over, and getting them very severely distorted in the process. I must say that. The situation is this: I don't have to hide anything, and I don't hide anything. Anyone can ask to come to Dounreay, and we will welcome them. They can see the process of operating reactors, they can see the process of the R & D, and they can ask us positive questions about what they've seen.

CHAIR It would be interesting to know, who's been to Dounreay here? Quite a few have. Well, I think that's very interesting.

BLUMFIELD No decision has been made, but several suggestions have been made.

CHAIR (Question from the floor). How many of your visitors to Dounreay would you think are potential terrorists or saboteurs?

BLUMFIELD I make sure that they do not sabotage us, that all the procedures necessary are carried out.

CHAIR You talk about the U.K. playing its full part in programmes of international collaboration. Could you specify the other countries with whom we are collaborating, or with whom you wish to collaborate?

BLUMFIELD We have a collaborative agreement with the Americans, also with the French, the Russians, the Germans and with the Japanese. On safety there is a complete open interchange between countries.

KEARTON There was a suggestion that we might buy C.F.R. technology from elsewhere. In Fast Reactor work, are we level with, slightly behind, or slightly ahead of other nations?

BLUMFIELD As far as reactor technology is concerned we're ahead in many ways. The boiler system in the Prototype Fast Reactor is a real boiler; other people have ducked that issue. Other people have been using uranium and not plutonium; they've been avoiding the issues. We have dealt with the total fuel cycle for this reactor, and there is no other country in the world that's dealing with that.

CHAIR (Question from the floor). Can you indicate whether the cost of cleaning up sites used for nuclear processing and waste storage in Britain is included in the cost you quote for nuclear generated electricity?

BLUMFIELD The answer to that is that the waste storage is included in the re-processing cost. The C.F.R. is designed so that you can take it back to green field state with a minimum of difficulty compared with any other reactor in the world.

CHAIR (Question from the floor). Has any site for CFR-1 been chosen even provisionally?

BLUMFIELD No decision has been made, but several suggestions have been made.

CHAIR (Question from the floor). How does the United Kingdom programme differ from that of other overseas countries?

BLUMFIELD We are dealing with the total field cycle, and we're the only country that is so doing.

WHYTE Is the economic argument the determining argument in deciding between alternative sources of energy?

*Mr Nicholson, who replied for U.K.A.E.A. on economic matters answered this question as follows:*

NICHOLSON No, I do not think that straight economics alone determine the issue. In all the policy decisions there are areas of uncertainty to do with future supplies and future prices, with amenity, and risk, and the rest of it, which are all taken into consideration in major decisions of this type. There are various means by which economists do just that.

NICHOLSON Well, first of all the A.E.A.'s reference programme was made in 1975 for our R &D planning purposes, and is now about two years adrift. If we now postulate a start for the CFR-1 which is considerably later, we could not build up to 33 gigawatts by the year 2000. That would require ordering of about 3 gigawatts a year through the 90's and that would not now be achieved.

CHAIR We have been dealing with the year 2000 as the reference year. Am I right in thinking that the Fast Breeder Reactor programme is likely to be of little assistance for energy problems before the year 2000?

NICHOLSON It would make little difference in terms of the total amount of power coming from Fast Reactors in the U.K. as a percentage of our total electrical system before the year 2000.

COTTRELL When do you think that it would be possible for this country to have the 33 gigawatts of Fast Breeder power? By what date?

NICHOLSON Between 2005 and 2010.

SEARLE If you have a problem of long life wastes and they're going to have to be looked after for a long time, how do you work out in present day prices how much that's going to cost?

NICHOLSON The method used for assessing the cost is to assume a certain cost for looking after these wastes for a very long time, and to postulate an investment—as it were, the annuity which would pay for the operating costs of looking after those wastes.

SEARLE Isn't it really the truth that generation after generation will be paying for this storage?

NICHOLSON Your statement is correct.

CHAIR (Question from the floor). The Department of Energy itself has quoted one thousand million pounds as the cost for the Steam Generator Heavy Water Reactor. Would the CFR-1 not cost as much?

NICHOLSON A thousand million pounds— a billion—is a satisfactory figure to use at present prices for the cost of CFR-1, with 1,300 megawatts. This would include manufacturing development and some investment in associated fuel plant.

CHAIR (Question from the floor). Mr Surrey and Dr Dombey

made the point that it might be preferable to build a 250 megawatt CFR instead of a 1300 CFR. Would you from the comment on this suggestion?

BLUMFIELD Obviously a larger sized reactor will be cheaper per kilowatt that a smaller sized reactor. We also have to take into account the C.E.G.B.'s own technical programme. The C.E.G.B. would prefer to go to a multiple of the 600 megawatt generator, which gives you 1200 to 1300 megawatts.

*Sir Montague Finniston responded to an invitation to make a submission:*

The problem areas in building the CFR-1—design, engineering, materials, safety etc.—can be identified and the U.K.A.E.A. research and development facilities supported by the industries which can serve it and by extensive interchange of information and experience with other countries with advanced nuclear technotloies should be capable of providing workable solutions. There is no reason to suppose that present "ignorances" are of sufficient moment to argue against approving the CFR-1.

Although the reactor is intended as a commercial prototype it need not necessarily, and most probably will not, incorporate all the economic refinements to provide lowest cost power; on the contrary there will be a proper tendency to conservative designs, the guiding criteria being reliability and safety in operation. The situation is not unlike that which obtained in the construction of the first thermal reactors (whether for research or power).

The degree of confidence engendered by the construction and operation of a single unit tempered by reigning political/financial/social considerations at the time will be the determining factors to whether or not there will be a proliferation of commercial fast reactors.

The experience with British engineering industry has not been entirely satisfactory when first put to the test for the nuclear industry, but with time and effort the services from the engineering industry has improved. The U.K. has in-house capability to manufacture or fabricate the majority, if not all, the structures required in the fast reactor; the lessons of the past 25-30 years in industrial nuclear technology has been to up-grade the engineering standards of products in common use, e.g. boilers, pumps, etc. to the higher levels of quality, performance and reliability demanded of nuclear applications. Where British industry will not, there is always foreign manufacture to fill in.

One of the more important parts of the reactor is the fuel element. This would be the responsibility of the U.K.A.E.A. and there is a degree of expertise which is the equal of any other fuel element manufacturer in the world.

The delays in completion of nuclear reactors in the past have been primarily due to factors not connected with nuclear technology per se but with the industrial temper of the country in works, on site etc. There have been technical difficulties which have emerged during construction of commercial reactors; these have not, however, affected the safety of the reactor but its expected performance (in thermal reactors) and in starting dates (in fast reactors). The experience gained in the P.F.R. and developments since then should stand in good stead in the construction of a commercial fast reactor.

Sir Montague Finniston replied to questions as follows:
CHAIR Your evidence was frank about the problems that have been encountered in engineering and technology in the past. Would you explain the grounds of your more sanguine feelings about the future?
FINNISTON I did not think I was sanguine. You mis-read the evidence, I'm sorry. I just think we're more aware of the problems of building a commercial reactor on which one could rely. The standards of engineering for which the first reactors in the U.K. were initially catered were not high enough or sufficiently sophisticated; and further there has just been a report published about the construction industry by the National Economic Development Office that shows the parlous state into which this particular industry has fallen in performance. I was trying to say that British industry—the construction industry in particular but the engineering industry generally— has failed to deliver, has failed to perform satisfactorily, has failed to give good value for money in the past, although the potential for doing well was all there.
COTTRELL From your practical experience would you advocate buying the Fast Breeder technology on license from overseas, as a practical policy?
FINNISTON This particular area of very advanced technology is highly specialised, sophisticated, subtle and requires a great deal of expertise most of which is still to be garnered. I do not see this totality of information, data and experience available anywhere else in the world. In such circumstances I fail to see where we can rely on the rest of the world for patents or licences sufficient to advance our programme. And in practical experience and skills we could match anybody else.
CHAIR We have just been told that by 2005 CFRS should be fully operative at 33 gigawatts. With your long knowledge of British industry, do you think that these dates can be kept, or are there likely to be delays?
FINNISTON Oh, I think there are going to be delays. Since it is going to be a Government programme you can be sure that they are not going to come to a decision quickly, and certainly not to a decision which encompasses 33 gigawatts. No, 2005 or 2010 is not too far ahead.
WHYTE How much of a blow would it be to the engineering

industry if the project were abandoned? Would it have serious effects?

FINNISTON I think you would have to look at who was going to benefit from the results of the developments arising from investment in the CFR-1 which demands specialised heat exchangers and specialised engineering not generally part of the industrial world. I do not think the boiler makers or turbo-generator manufacturers for example would lose a great deal. The pump manufacturers? Again I think the same conclusion would apply. That leaves then the reactor proper. There would be the vessels. There are many chemical plants now which are dangerous and have to have vessels developed; that technology could be learned on that relationship. My answer to you would be that it would be a blow but it would not be a blow that would be likely to reduce the industry's capability to meet other industrial demands in the technical sense.

The Nuclear Power Company made the following submission of the technological feasibility of CFR-1:

We consider that the engineering feasibility of fast reactors has been demonstrated by years of actual operation in several countries. The operation of Phoenix in France and P.F.R. at Dounreay should increasingly confirm the feasibility of fast reactors which are broadly similar in design to a full-scale demonstration unit. This will also be true for the fuel cycle operations—the fabrication of plutonium—containing fuel, and the reprocessing of such fuel after irradiation. In our view a large amount of work is nevertheless required to establish satisfactory levels of operating reliability for the major components of a full-size industrially acceptable fast reactor. The work must be undertaken in detail on real components in realistic environments. It must have the breadth to be representative of future artifacts. In the field of reliability some of the key issues are the manufacture of leak-free boilers, the integrity of structures after long exposure to fairly high temperatures and the integrity of fuel elements.

In the field of safety, one principal task is to establish data and information systems so that small departures from satisfactory operation cannot deteriorate to produce larger damaging effects undetected by the station staff. Another is to identify all faults, even though they are most unlikely to occur, which could lead to escalating damage within the reactor, and providing assurance and safeguards against those faults. A third task is to protect the public against those faults which conceivably could escape through the previous two safeguard systems. A very large amount of progress has been made on all these topics in all countries having large fast reactor research and development activities (Russia, U.S.A., France, Germany, Japan and Italy, as well as the U.K.) There is presently a large amount of international co-operation in the field of safety research.

This stage of development and demonstration requires a sufficient period to prove reliability and to afford confidence in the adequacy of knowledge and understanding. It is not to be thought that no new aspects of long term behaviour can arise. On the other hand, it does seem likely that the experience of the past, and that from Phoenix and P.F.R. during the next few years, will make it possible and responsible to undertake a full-scale demonstration programme of two or three fast reactors in Western Europe coming on power sequentially during the 1980's. Such a programme would afford confidence for the construction of fast reactors as a significant component of the Western European electrical systems in the 1990's. If such programmes are required it will be because fossil fuels and uranium are both scarce and expensive. Credible cost differences between thermal and fast reactors are unlikely to be important in affecting the economic advantage of the fast reactor. Reliability and safety, coupled with the social acceptability which we hope will result from the envisaged demonstration, will be decisive. It is for this reason that we consider the foregoing programme to be necessary. Whether it will be judged adequate depends upon the experience over the next decade and the prospective alternatives.

Mr Wolff gave evidence on behalf of the Nuclear Power Co.:-

COTTRELL I'd like to hear your views on the time scale in moving into the ultimate production phase, which is the phase that would concern your company. I'd also like to ask about the capacity of your industry—how many such reactors do you think your industry could build simultaneously?

WOLFF First of all, on the time scale of CFR-1, I agree with Mr Blumfield's estimate that if we had a decision now, it would take three years to get organised on the modest scale necessary to start work on CFR-1. The production phase might then begin around 1990. As for the capacity of the industry, let me take as a reasonable supposition that we might require 33 gigawatts (Dr Nicholson's figure) to be installed and working in 2010. We consider a ten year period not unreasonable for building up industrial capacity for quantity production. That means, over that period of time, working up to about 5000MW per year. That is certainly within the sort of capacity that the U.K. power plant industry has achieved in the past. And providing that preparatory steps are taken, I can't see any particular reason why it should not be achieved again in the future, even if today demand is very low.

SEARLE Why do you need to go ahead now if there will be a 3 year time lag between now and the commencement of construction?

WOLFF The go-ahead now will allow design to pin-point on something which can be built and will allow the specific work to be undertaken which is still required on components and component development which we would not think it would be very prudent to start on site.

CHAIR (Question fron the floor). How many thermal nuclear reactors would be required to provide plutonium for the breeders?

WOLFF If we're talking about water reactors then our estimate in N.N.C. is that you have to run a thermal reactor for about 10 years to produce the total inventories for a fast reactor of the same output.

CHAIR But you've a large store of the stuff already?

WOLFF Yes, there is a large store and the Magnox reactors which are providing the bulk of our nuclear electricity are much better plutonium producers and I think they take between 4 and 5 years to produce plutonium for fast reactors.

CHAIR (Question from the floor). If you don't get a decision soon about CFR-1, will you have to disperse your team?

WOLFF I don't think that is a question that can be answered just like that. If we don't get a decision to start constructing CFR-1 in 3 years' time, we could continue work according to a less intensive plan. If I may turn the question, if there were a decision today to proceed with CFR-1, we should have very substantially to increase the numbers of our staff that we have got on the CFR-1 project.

HENDERSON Would you comment on the sentence in your evidence: "Credible cost differences between thermal and fast reactors are unlikely to be important in affecting the economic advantage of the fast reactor?" I had the impression that the question of cost advantage was really rather decisive in making an economic assessment between fast reactors and thermal reactors. I thought it was generally agreed that the capital cost of a fast reactor would be higher but that the fuel cycle cost ought to be considerably lower. A great deal therefore depends on how much more the capital costs are likely to be and on what the price of uranium will be. I must confess I haven't seen any evidence that has really convinced me that a Commercial Fast Reactor will pay. Can you help me?

WOLFF 'Credible cost difference' refers to credible capital cost differences. We are admitting—as my colleagues in U.K.A.E.A. have said—that we don't today know precisely what a fast reactor is going to cost. We certainly know from experience that even CFR-1 is not going to give you the true capital cost of the Commercial Reactors which we may be building at the rate of some 5,000 megawatts per year.

However, we have done a number of cost sensitivity studies quite independently of the Authority. In order to clarify the issues, we have done these on a cash flow basis. If you don't do that, then your very early costs, particularly development costs, entirely swamp the later scene. (If you use the Treasury recommended discount rate of 10% you may think that hardly matters what you're doing 15 years from now. What I'm now saying is contradicting some things which were said to this meeting earlier in the morning. We considered in 1975 money terms uranium prices going up to 120 dollars per pound. Whether that is a true ceiling or not I don't think is known, but it is a guesstimate of the cost level of extraction from the really large resources of uranium available in the sea. That's just to establish a cost ceiling and it's not taking into account that uranium might physically be unavailable. If you start talking about that sort of figure (and it seems to us not improbable if you take a 50 year view), then you could easily afford to pay even twice as much for the reactor portion of a fast reactor as for the reactor portion of a thermal reactor. The sort of figures that Mr. Nicholson was giving you this morning still hold good. Hence we forecast that you will be getting very much cheaper generation than by the competing thermal reactor of that time. I refuse to guess by how much large a factor even thermal reactor generation is going to be cheaper than generation from fossil fuels.

WHYTE What eventuality do you think would justify stopping at CFR-1 and not going on to CFR-2; or once we have started CFR-1, are we on an inevitable progression?

WOLFF We take a stand that there's nothing inevitable about what happens after CFR-1. However, the basis for starting CFR-1 will be that it is highly likely that criteria of safety, economy and reliable operation can be met.

CHAIR (Question from the floor). Would CFR-1 be sited with its own reprocessing plant as recommended by the Royal Commission on Pollution in the Environment?

WOLFF I'm unable to say.

CHAIR (Question from the floor). Could the nuclear industry also manage to build the necessary thermal reactors to support the Fast Reactor programme in its early stages?

WOLFF At the present level of capacity in this country we can build far more than we need of almost anything you can mention in the power plant field. To-day's problem concerns how much of present capacity will have to close down because it cannot be sustained in the short term.

---

1I. C. Bupp and J. C. Derian, Technology Review (M.I.T.), July/August 1974.

# *Chapter V*

# CFR-1 and Alternative Sources of Energy

The intention underlying this session was to determine the potentialities of alternative energy sources, and to consider the extent to which investment in CFR-1 might preclude their full development and might have undesirable international implications.

## Coal

The National Coal Board drew attention to their PLAN 2000 and Mr. M. J. Parker, Director of the Central Planning Unit gave evidence as follows:
CHAIR How much coal can we expect to get out of the ground per year and in particular in the year 2000?
PARKER We have to start with an understanding of the current situation in the industry. I refer not to industrial relations, but to the physical situation of the industry and its current inheritance. The average age of our pits is 80 years, the average distance from the shaft is several miles, and therefore we have in considering our future capability to take account of the fact that we have to replace before we get net expansion. We have planned to build up the deep-mined capability from the present level of about 120 million tons capacity up to about 150 million tons capacity deep mined by the end of the century. To do that we shall have to provide something like 4 m. tons of new capacity per annum, since we have not only to create new capacity for growth but also to create new capacity to replace the Victorian capacity which will be exhausted progressively during the next 25 years. In addition to the deep-mined coal we are seeking to maintain broadly the present target of 15 to 20 m. tons of open cast surface mining for which there are sufficient surface reserves to sustain it to the end of the century. We are looking at an ultimate capability by the end of the century of something like 170 m. tons of British coal mine output compared with the current level of production of around about 120 m. tons. This estimate is not made unconditionally, but on the assumption that we get all the capital involved to carry this out and also that we are able to obtain the productivities necessary with this new capacity to make them economically attractive. So the name of the game

is still economics. We plan to provide indigenous sources at a price competitive with the world energy price. Our plans are not as the Americans would say predicated upon the total failure of the nuclear industry in this country. They presuppose that there will be a continuing nuclear expansion, but perhaps not as rapid as some parties have advocated.

CHAIR Is it usually a presumption to say that this is not part of your planning?

PARKER It is part of our planning in relation to potential demand for our product and therefore that does condition the degree of our physical expansion.

CHAIR If there wasn't a nuclear programme could you produce more?

PARKER I think that's unlikely for the reasons I've tried to indicate in regard to our inheritance. We have a problem of bringing on sufficient new capacity in the time available. It may conceivably be possible, but it would require a degree of national commitment which is not in sight at the present time.

CHAIR So the assumption is irrelevant to the end product?

PARKER Not necessarily because it would condition the commitment to resources in the intermediate period.

COTTRELL Are the figures you've been describing what we might call conventional foreseeable methods?

PARKER That is correct.

COTTRELL What are the prospects for new methods of getting the energy of the coal out of the ground?

PARKER We have looked at this in some detail; our view is that this is not likely to make a significant impact during the next 20 years or so, though we don't rule it out for the period beyond then and this again is an economic question—what cost is one prepared to incur to find out new methods in relation to the value of the product recovered? The most hopeful present non-conventional method is underground gassification. We've done a world-wide survey of this and our view is that the world energy price has still got some way to go before this is going to be economic. With present techniques there's a very poor recovery of the coal in situ. Also the underground combustion is often very incomplete so that it's difficult to continue this process with faults and water and so on.

WHYTE One of the effects of the Victorian enterprise was that it had really very bad effects upon the environment. Would the projected new developments have the same effects or would they be better?

PARKER We would do our best to ensure that they were better. I think it was a feature of Victorian mining that nobody had then invented the environment! The main characteristic of Victorian coal mining was its intensiveness in terms of the numbers of shafts and the numbers of entry points to the seam. We are moving towards much greater concentrations of production and therefore much fewer numbers of entry points, and we think that environmentally this will be better.

CHAIR (Question from the floor). Can you say how many miners will lose their lives because of the planned expansion of coal?

PARKER No, I can't.

CHAIR Do you think more miners will lose their lives?

PARKER I wouldn't have thought so.

CHAIR (Question from the floor). How much change has there been in coal mining research and development since the oil crisis of '73?

PARKER We're doing about three times as much.

CHAIR You'd get up to 170 m. tons of coal by the year 2000 in your forward planning. For how long can that be sustained in your estimate of reserves? Can it be increased, and if you plan to increase it, at what rate?

PARKER We haven't looked that far ahead with sufficient detail to give a definitive statement. We know that in principle there are 300 years at current rates of production. How we are able to exploit that will depend upon technology and economics.

## Solar and Wind Power

The potentialities of solar and wind power were described by Dr. J. C. McVeigh, Head of the Department of Mechanical and Production Engineering at Brighton Polytechnic and Vice-chairman of the U.K. section of the International Solar Energy Society. *His submission on solar power was as follows:*

Some countries may be able to provide nearly all their energy from solar systems, but in the U.K., where we have about half the solar radiation of the U.S.A. or Australia, we may only be able to supply between 10 and 20% of our energy requirements in the long-term—say 2020. Although small by comparison with what might be possible in other countries, it would be a valuable and inexhaustible source. Each year the world's solar energy income at ground level is about $3 \times 10^{24}$ joules. This figure may be compared with the present proved world fossil fuel reserves at around $0.04 \times 10^{24}$ joules. These fossil fuel reserves are equivalent to about one week's supply of solar energy landing on the earth.

Substantially more attention should be given in the U.K. to the study and development of solar energy application as a significant long range issue in national energy policy. Existing knowledge and expertise in the U.K. is substantial, so that a good base exists from which to find a rapid expansion of R and D for use within the U.K.

Highest priority should be given to achieving more cost effective systems of solar collection through well financed development programmes by industry rather than financing further basic research. The A.C.O.R.D. document points out that solar energy has the potential to make a significant impact in space and water heating, but nevertheless its figure of 2% of the total energy demand by the year 2000 seems rather low. Limitations on the use of increased solar energy include the seasonal mis-match between solar input and maximum energy demand and the relatively high cost of domestic solar water heaters. The British Standards Institution is now establishing a technical committee to prepare standards for solar water heaters and there is evidence that some large companies will enter the solar water heater market. This should reduce the unit cost of solar water heaters.

A figure of about 6% of total demand could be satisfied by solar energy by the year 2000 under the following headings:

| | |
|---|---|
| (a) domestic water heating for existing housing | 1% |
| (b) space and water heating for new housing | 2% |
| (c) low grade industrial heating application | 2% |
| (d) commercial water and space heating | 1% |

A solar cell converts some light directly into electricity and has many attractions as a potential source of electrical power for homes, offices, hospitals and factories. It has no moving parts, has a long life and requires little or no maintenance. It works quite satisfactorily in diffuse daylight from a cloudy sky. Even at the present very high capital cost, it is economically viable in an extensive range of small power applications. If expectations of radical cost reductions are realised it could become widely used in the future.

*Dr. J. C. McVeigh's submission on wind power was as follows:*

A zero growth energy policy would be achieved in the type of society outlined, for example, in Thring's "Man, Machines and Tomorrow" and Schumacher's "Small is Beautiful". If we are prepared to accept this pattern of energy consumption, there is the possibility that wind-power alone could provide at least 25% of the total U.K. energy demand, provided the necessary research effort is employed to investigate large-scale high-grade energy storage.

The contribution which wind-power could make in the U.K. has been under-estimated in the A.C.O.R.D. document which is based on the concept of a small number of large, expensive, high technology machines on a limited number of windy hill-top sites. Smaller windmills, with blade diameters of less than 50 metres, can be developed very rapidly as the necessary medium technology is already available. In mass production these windmills would be comparatively inexpensive and well under the present capital cost per installed kilowatt of nuclear power. They could be sited anywhere in the country where wind conditions are adequate. Some of the advantages of wind-power in the short term are as follows:

1. Wind energy is high grade energy. It can be used as mechanical or

electrical power or de-graded into low temperature heat.

2. In temperate and cold climatic zones the availability of wind-power broadly matches the seasonal demand.

3. Wind-power can be used with existing systems and does not require changes to life-style or facilities.

4. No new fundamental technology is required for the horizontal axis machines, as development can draw heavily on existing knowledge in allied fields.

5. About 80% of typical modern power plant consists of conventional equipment. The development costs of the remaining parts, mainly the windmill rotor and the associated control equipment, will be relatively low.

6. Because of the moderate economy of scale, wind-power units will be relatively inexpensive in production.

7. Export potential for wind-power plant is substantial.

8. Of all the renewable energy possibilities considered in the A.C.O.R.D. document, wind-power is the only one which could provide the most rapid, the least capital intensive and most crisis-proof replacement for imported fuel.

9. Wind-power has no known environmental hazards. Although the visual impact has been stressed by some critics, the architectural challenge of modern technology windmills can be met and there are pleasing examples both on paper and *in situ*. In some cases one modern windmill tower may replace several transmission pylons.

10. In the outer regions of the U.K. there is a direct relationship between current electricity costs and the strength of the wind. Wind-powered domestic and district energy systems and some agricultural and desalination systems could be provided for the islands of Scotland, the Isles of Scilly, the Isle of Man, the Channel Islands at the cost to the consumer of less than is currently being paid within a few years.

(Acknowledgements to J. S. Shapiro and G. W. Pontin for access to recent technical notes).

Discussion of these two alternative energy sources began with a series of questions to Dr. McVeigh about solar power.

COTTRELL You say that highest priority should be given to achieving more cost effective systems of solar collection, but I think as regards solar collection for domestic heating the scientific part of the work is very largely done, in that industry has already taken up the problem and there are several industrial firms who are already pursuing this commercially. In view of this, what do you mean by highest priority? If a firm sees an opportunity surely it will follow it with its resources? I believe many are doing so.

McVEIGH It's a question of accelerating the rate of this progress. I would agree entirely that very little fundamental work is needed on the simple flat plate collectors for low grade heat collection but far more emphasis is needed on the higher technology collectors to produce high temperatures, and this is where one could see export markets developing. Also we do need to think rather more about what is potentially available; my estimate of 1% for domestic water heating from the existing housing is made on a very modest assumption that only a quarter of existing houses will have solar water heaters by the year 2000, and if industry accelerates now and does the necessary work to take this up for mass production it will need investment and possibly government incentives in the way of taxation relief, as is being done in other countries.

SEARLE You say that in the U.K. we may only be able to supply between 10 and 20% of our energy requirements by about 2020. Is that the percentage of our current demand, or is it the projected demand up to the year 2020? What sort of resource and energy input is required in order to achieve those energy savings?

McVEIGH The figures were current when I prepared this submission. They are fairly conservative because they don't take into account the enormous contribution one could get from photobiology if more money were put into research in its research. When I said "highest priority should be given to the more cost effective systems," I was thinking not only of the engineering of flat plate water collectors but also photobiology and photovoltaics. My percentages were broadly based on today's demands.

SEARLE Where solar panels are concerned, what sort of energy investment does one have to make in terms of the energy output one derives?

McVEIGH It depends on what sort of basis you work. It would not take more than 4 years for a flat plate collector to repay its energy costs. For photovoltaics a rather different figure emerges at present, but the long term projections for photovoltaics are extremely optimistic in that people talk in terms of a factor of 100 in cost reduction. One could envisage them as integral roof panels that were photovoltaic collectors. This is being studied in the United States at present.

CHAIR I live in a smallish detached house on Wandsworth Common. Would solar heat be economically feasible for me?

McVEIGH You could put in solar water heaters and expect to save about 1500 kilowatt hours of units of electricity. Four square metres of panel commercially installed would cost you about £500-600.

WHYTE Do you feel that in the present circumstances enough government money is being spent on this kind of research?

McVEIGH It only needs about £5 million to develop a 46 metre diameter windmill, which is a one megawatt unit. I don't think enough money is being spent on the alternative sources; there is rather more conviction in government circles about the whole field of solar energy. I hope that the scale will increase from the present half million pounds by a factor of at least 4.

CHAIR (Question from the floor). Sir John Hill has talked of selling nuclear reactors to the third world. Would they do better there with solar energy?

McVEIGH The third world is extremely fortunate in that they've got a very much better solar climate than we have. We can "collect well" in this country for the three months on either side of mid-summer. Most of the third world countries have got very good and abundant sources of solar energy, and other countries have noticed this, and they are putting a lot of government funds into research and development, particularly the development of solar powered water pumps for irrigation.

SEARLE You've told us in connection with solar power that by the year 2020 about 20% of our current energy demands will be met by that source. Also in your submission on wind power you say that wind power alone could provide at least 25% of the total U.K. energy demand—presumably the current demand. We know that both are seasonal in the contributions they can make. Do you envisage local marriages of the two systems so that you will have within a locality both systems operating, thus complementing each other over a period of 12 months?

McVEIGH Yes. The other important point that I've made is the question of energy storage. We're talking about high grade energy storage on various scales, that is, the individual scale, the district scale, and the national scale, and various means of storage mechanical, electrical, high temperature, as well as low grade storage. It is possible now to buy an off-the-shelf chemical storage system for a house in Germany—the Laing Institute have developed one. They're also working on high temperature high grade underground storage.

SEARLE You're saying that 45% of our current energy demands can be met by those sources?

McVEIGH Yes, if you throw in wind and solar together. As far as official government predictions are concerned it is assumed that a few high technology windmills will be placed on high grade sites with very good wind conditions. But if you go in for lots of smaller machines you can get the benefits of mass production. The economics have been very carefully worked out and are contrary to the views of the C.E.G.B. and the Atomic Energy Authority. The typical cost that I'm talking about in mass production is of the order of a penny per kilowatt hour to the user.

ADAMS What do you mean when you say that windmills can be sited anywhere in the country where wind conditions are adequate?

McVEIGH I mean that you start off by using the more wind intensive high energy sites off the north west coast. But you can go down to a wind speed of 6 metres per second or less which applies to about 25% of the country. You get the benefits of scale economies as you build more and more windmills, and you build up experience; also you can get them into local housing with architectural features such as a wind wall. This is really just a large box with a windmill in the centre pointing in a fixed direction.

CHAIR (Question from the floor). How many acres would be covered by sufficient 1 megawatt windmills to be equal in output to a 1300 megawatt commercial fast reactor?

McVEIGH You can site large windmills about 3 blade diameters apart, and you can group them in clusters.

CHAIR How many would be needed outside my Wandsworth Common house in order to produce 25% of energy demand?

McVEIGH I don't think we'd be building them in cities like London in the year 2000.

HENDERSON If the prospects that you hold out are really so attractive why are firms not already trying to seize the opportunities and going ahead developing systems without government support?

McVEIGH It's not only firms. If we look at the rest of the world we find the United States, Sweden, Denmark, Holland, Canada, Israel, West Germany and Japan have plans now for wind driven electrical generators to be linked to the national grid. We haven't. There are signs that large companies are getting interested and even that the government is beginning to take notice—but too slowly!

CHAIR Would the visual impact be no worse than that of television aerials?

McVEIGH I think they'd be more attractive than television aerials. Your landscape is already covered with thousands upon thousands of electrical pylons!

HENDERSON Could a wind generator replace pylons completely and not need any back-up?

McVEIGH Wind is a fluctuating source of energy and you do need a back-up. You can do this locally with small generators or by the year 2000. I'm quite convinced that if the money's put into research (such as in the Laing Institute in Germany), one would see large scale energy storage capable of coping with the loads when there was no wind. Statistically we never have more than five consecutive calm days in the U.K.

CHAIR I haven't invited anyone to speak about geothermal sources because Energy Paper No. 9 of the Dept. of Energy summarizes the prospect. Its first conclusion is that "prospects of geothermal development in the U.K. appear to be sufficiently favourable to warrant more detailed study in certain locations." And until that has been done there doesn't seem to be any point in asking a witness to come forward. As for tidal energy we hear the Secretary of State for Energy telling us that it is not economically viable. H.M.S.O. have published Energy Paper No. 2 on *Methane derived alcohols,* but I was unable to find anyone who could speak authoritatively in economic terms about methane from organic sources. (Since the Hearings took place, I have been introduced to *Methane: Planning a Digester,* by P. J. Meynell (Prism Press, 1976).)

**Wave Power**

*The submission on wave power was by Mr. S. H. Salter from the School of Engineering Science at Edinburgh University:*

A great deal of energy is brought to our shores by the waves of the North Atlantic. Mollison et al. have analysed data from weather ships collected by the Institute of Oceanographic Sciences. The average power density through the year is 90 kilowatts per metre. The seasonal breakdown over a thirteen year period was as follows:

| | |
|---|---|
| Winter (Dec to Feb) | 177 kW/m |
| Spring | 78 kW/m |
| Summer | 26 kW/m |
| Autumn | 91 kW/m |

*Reliability*

Maintenance of equipment at sea is difficult. We will therefore have to achieve overall reliability by modular construction, multiple connections and elaborate fault isolation techniques.

But, as far as the waves themselves are concerned, 'firmness' is remarkable. For example, in winter the power density in the Atlantic is more than 50 kilowatts per metre for 80% of the time. The nuclear stations of the C.E.G.B. are able on average to supply power for 70% of the time. For the whole year wave power exceeds 40 kilowatts per metre for 55% of the time. This matches the figure for the twenty most efficient fossil fuel stations. It seems that the waves themselves do not set an unacceptably low limit.

*Extraction devices*

The Department of Energy is supporting four serious studies of extraction mechanisms. The problems which have to be solved are formidable and their solution will call for a very large engineering effort. Wave forces in extreme weather conditions can be enormous. However, all the designs share the common factors of using familiar materials in well-tried conditions. I believe that the technology can be developed in the time left to us. Mass production could begin in 1985.

Practical wave power machines would extract a high proportion of the available power in calm conditions and progressively lower proportions in bad weather. Sensible mean annual yields willl probably lie between 20 and 50 kilowatts per metre. A low rating gives a high duty factor but can make no use of the high winter level. Conversely, a high power machine will be used in the summer. To optimise the design we have to decide on the relative values of summer and winter electricity, and whether we want the most energy or the lowest cost.

*Costs*

Predictions of future costs of electricity are extremely difficult to make because of factors like the interest to be charged during the development and construction phases. Such analyses as we have been able to do show that wave-generated electricity would not have been competitive with the very cheap Middle East oil prices of the 'sixties but that today we could expect a price of between one and two pence per kilowatt hour.

Several separate energy accounting studies have concluded that the energy repayment period is acceptably short provided that concrete is used for the bulk of the structure and that current performance curves are used. The results lie between six and eighteen months even if we do not include a premium for electrical as opposed to thermal energy.

*Sea space*

There are about 600 kilometres of suitable Atlantic sea front available to the United Kingdom and a further 600 round the Republic of Ireland. Stations in the Channel approaches would require rather better marine traffic discipline than now exists. Power in the North Sea declines from Atlantic levels as one moves south from Sea Area Viking. The costs of underwater electrical transmission are known and are high enough to prevent the use of distant offshore installations at present. Some chemical routes may prove to be possible.

*Implications–Technical and Environmental*

Extensive deployment of wave power installations would call for major rearrangements of the national grid with strong pressures for an increase of transmission voltages to the levels used in Russia and Canada. Hydro-electric plant and pumped storage would be particularly valuable and it would be necessary to continue the use of existing coal-fired stations.

Wave power is remarkably benign. It introduces no new chemicals or heat into the biosphere. There will be a very slight cooling of the water on the beaches—about a hundredth of a degree centigrade.

Installations capable of supplying the whole of the present United Kingdom electrical load would grow some hundreds of thousands of tons of mussels and barnacles each year. These would appear to constitute our principle waste disposal problem.

*References*

1. U.K. Energy Statistics 1976, Dept. of Energy, H.M.S.O., London.

2. D. Mollison, D. P. Buneman, S. H. Salter, Wave Power Availability in the N.E. Atlantic, *Nature,* Vol. 263, No. 5574, 223-6, 1976.
3. C.E.G.B. Statistical Year Book, 1975-6. C.E.G.B. London.
4. L. Draper, Extreme Conditions in British and Adjacent Waters, *Proc. 13th Coastal Engineering Conference,* July 1962, Vancouver, Canada.

Mr. Salter was questioned as follows:

CHAIR Are you getting as much money for research for your project as you think necessary for its speedy realisation?

SALTER Yes.

CHAIR My goodness, the first chap on research I've ever heard say that! Well, let it be counted for righteousness for the Secretary of State for Energy!

SEARLE One of the problems that have been evident in nuclear energy is that of extrapolating from an experimental beginning to a commercial situation. Does the same problem exist for wave power?

SALTER You could build 500 miles of installation as easily as you could build 1000 miles of motorway. There is a lot of concrete involved, and if you wanted to do the typical sort of load that we have on the electrical network, at the moment, you'd need to use most of the available sea space that we have. This is about 600 kilometres.

SEARLE What effect would that have on people going to the seaside?

SALTER They would be sufficiently far offshore not to be seen from the beach. They'd be about 3 or 5 miles offshore.

SEARLE And what about storms? Is it likely that we'd have great concrete things lashing up against the beaches?

SALTER No, we would try to avoid that because they would be destroyed if that happened. The forces in storms are formidable and that's our main problem.

WHYTE You refer in your evidence to the expectation of a price of between 1 and 2 pence per kilowatt hour as your present projection of costs, taking, I presume, 1976 values. What's the C.E.G.B. and South of Scotland Electricity Board's reaction to that figure? Do you think you're being a bit optimistic or are they prepared to agree?

SALTER I sent them the analysis which we used but they haven't commented on it yet.

COTTRELL The idea is scientifically feasible. Then comes technical feasibility, and after that will come economic feasibility. When could be expect to have a large power output from this system?

SALTER We need to do experiments for about another 18

months to have a fix on the final structural forces. We'd then need to build prototypes and test them thoroughly in winter. This takes us into the early 80's, perhaps 1983, and you would not be able to build them very rapidly until after 1985. At that point if you really were in a flap and were able to build in the sort of way that we built the liberty ships in the war, you could be taking the strain in the 90's.

CHAIR (Questions from the floor). Do you envisage wave energy being brought ashore by any means other than electricity?

SALTER The inshore energy would be electrical. We couldn't do offshore wave energy with electrical transmission; we'd have to find a chemical route.

CHAIR Experience of corrosion effects on deep sea structures is only just coming through from oil drilling structures. Can you make any realistic estimates in your field?

SALTER Concrete looks very good.

CHAIR You mean it'll last?

SALTER Yes.

CHAIR I don't think it looks good, but I take your point. Is there a possible benefit to our declining ship building industry from the construction of wave power installations?

SALTER I think employment is a very bad reason for making a technical decision.

CHAIR Yes, but is there a possible benefit to our declining ship-building industry from the construction of wave power installations?

SALTER Yes.

CHAIR Given current technology, is the conversion to 700 kilovolts from wave-power installations feasible? If not, will this technology be feasible by 2000?

SALTER I think it's feasible now.

**Nuclear Fusion**

*The following submission on nuclear fusion was made by Dr R. S. Pease, Director of the U.K.A.E.A's Culham Laboratory.*

Controlled nuclear fusion offers the prospect of generating heat and electricity from a very large and cheap source of nuclear fuel (lithium and heavy hydrogen) which has harmless reaction products (helium). Nonetheless fusion has some environmental impact including radioactivity; this has been the subject of both national and international studies whose results have been published.[1]

Unlike the various nuclear fission reactors—and in particular fast fission breeder reactors which have generated electricity on a significant scale for many years—controlled nuclear fusion is still in the research stage, and has generated no heat or electricity so far. In planning something as important to the nation as future electricity production, we can rely only on those methods which have been demonstrated to work. Therefore controlled nuclear fusion, however attractive may appear its potential advantages, cannot yet be considered as an alternative to proven methods of generating electricity. This situation is likely to persist until the 1990s at the earliest.

Power-producing nuclear fusion depends in the first place on achieving control and isolation of matter heated to temperatures of 100m°C. As I have outlined elsewhere ("The Potential of controlled Nuclear Fusion", Review Paper given at the British Association for the Advancement of Science Meeting, Lancaster, September 1976), it is presently routine to achieve temperatures of 10-20M°C; and there are good prospects of achieving the high temperature conditions required for a power producer in the next decade. Thereafter, however, extensive engineering and technological developments have to be applied if electricity production is to be achieved.

The most explicit programme planning so far published, that of the U.S. Energy Research and Development Administration, indicates plans which, if put into effect and technically successful, would result in prototype systems generating electricity in the later 1990s.

The United Kingdom programme of research on controlled nuclear fusion is carried out in a co-ordinated programme within the European Community. There is a high degree of commitment to this co-operation continuing through to a fusion reactor. However, the problem of getting programmes agreed by all the countries concerned, the relatively low level of proposed funding, and the technical difficulty of currently envisaged fusion reactors suggest that no significant amount of electricity will be generated by European prototype reactors before 2000 A.D.

Dr. Pease was questioned as follows:

McKNIGHT Can you give us any idea of the health and safety hazards associated with a fusion process? Also, can tritium be diverted for hydrogen bomb production?

PEASE As far as the first question is concerned we have published two reports on fusion reactors and the environment. The second of these was published by the I.A.E.A. on the advice of a Committee of which I was Chairman. The reports deal in some detail with your question and I hesitate to give a quick answer, but the advantages of fusion are plentiful cheap fuel with no reaction products which are toxic or dangerous. But we are anxious about the most accessible form of fusion, that involving tritium where there will be some radioactivity both from the tritium itself and from the structure. As regards the diversion of tritium to hydrogen bombs, I don't think this is likely. I'm not a weapons expert, so you'd have to ask somebody who is.

WHYTE Unlike Dr. Salter you appear to be dissatisfied with

the level of funding of your research. Can you suggest why it has not attracted more attention? Is it because it's in more direct competition with nuclear fission?

PEASE I ought to be careful how I use the words satisfied and dissatisfied. I'm a servant of the taxpayer and do the best I can within the resources made available. But I'm naturally worried and distressed when one sees that the Americans and to some extent the Russians are funding this line a good deal more strongly than we are.

CHAIR Do you think that by the year 2076 the commercial fast reactor, if we have it, will seem quite out of date and that we shall by that time have a fusion economy?

PEASE The easy answer for me is yes, but I would not wish you to make now a major policy decision for the year 2000 which is based on that opinion. In my view in the matter of fusion we will have to wait and see. I think we shall have to wait till about 1990 before that question can be answered in more detail.

McKNIGHT You refer to the greater support of the United States and the Soviet Union for fusion research. Do you gain as much by joint western European efforts?

PEASE The problems of fusion are difficult. The problems of uniting Europe are difficult. We're trying to deal with them both and it does take a little time to solve them.

CHAIR (Question from the floor). Fusion can be used to generate plutonium by using a uranium blanket. How will we avoid nuclear proliferation problems? Are you considering fusion fission systems i.e. breeding plutonium and burning it in fission reactors?

PEASE We are not, but it is a topic which is raised from time to time and there might be advantages in an energy system based on such a process. My own personal view here is that we must make the fusion reactor work first. Then we can give proper attention to these possible applications.

SEARLE Is your optimism with regard to dates shared by most of your colleagues? You say that no significant amount of electricity will be generated by prototype fusion reactors before 2000. Isn't that a very short period?

PEASE I didn't say it *would* be generated by the year 2000. I thought that my statement was fairly conventional, but I'm bound to point out that the United States Energy Research and Development Administration have published plans to produce electricity generating prototypes by the late 1990's. The misunderstanding centres on the word "significant". If a 500 MW (E) fusion power station had operated succesfully for a year or so by the year 2000, this would be highly significant of the potential of fusion. But the actual amount of energy generated would be insignificant. In so far as this Inquiry is concerned with generating power in 2000 AD, the prospects that fusion can make a big contribution by then are so poor that, as said in my evidence, you should ignore them.

COTTRELL In your Paper you seem fairly satisfied about the scientific prospects of nuclear fusion but you still have considerable doubts about the engineering and technological developments. I would like to know whether in your opinion that the engineering and technological developments will be solved? Can we expect to have commercial nuclear fusion within the next 50 years?

PEASE My personal reply to that is yes. But as far as making a major decision is concerned on existing investment in power systems that work, I don't actually see that this has a very great impact. As for the engineering problems there are two issues. The first is that we need to control matter at 100 million degrees. Secondly, even if we solve all the physics problems there will be a great many engineering problems which we have to solve and which are comparable with those faced by other big electricity generating systems.

CHAIR If the European Economic Community finally decides where it will locate its fusion research, and plans to go ahead, do you think there will then be adequate resources for the development of fusion?

PEASE It will make an enormous difference to have this sense of purpose in the Western European programme, and I would hope that we would make enough progress to justify the investment which I think the subject deserves.

**Alternative Sources—General**

*The following submission was made by Mr. Gerald Leach of the International Institute for Environment and Development:*

Fast breeder reactors can supply very little energy before the year 2000. Yet by this date the U.K. must be well into a rapid and sustained transition away from oil and gas as principal fuels.[2] CFR-1 is therefore irrelevant to the major energy problem facing this country, while its high investment costs (and implicit commitment to electrification) could draw resources away from much more promising solutions.

Among these promising alternatives is a wide range of renewable energy sources *combined with* measures to improve the efficiency of energy use. On their own, renewable sources such as the sun, wind and waves have a substantial but not obviously very large supply potential in the U.K. But in combination with energy savings to exploit self-reinforcing (synergistic) effects, the potential is dramatic and forms one of the most exciting—and as yet under-explored and under-financed—technological challenges of our time.

For example, in the domestic sector insulation to standards that are now cost-effective can reduce energy needs by 15-25%. This reduces the required capacity and costs for heat pumps, passive solar devices (e.g. large south-facing windows), active solar devices (e.g. rooftop collectors), solar storage, etc., while many of these simple techniques—such as solar and heat pump combinations—have strong self-reinforcing effects which increase energy savings and sharply reduce unit capital and operating costs. A recent U.S. study shows that annual energy consumption for a typical all-electric home can be cut from 30 to 5 MWhe by such means.[3] Reductions of the same order seem possible in the rest of the energy economy (with the exception of transport and most industrial processes) since over 70% of energy demand in these areas is for low grade heat. This demand is energetically well suited to combined renewable/conservation systems, though not to electricity production. For transport there are good prospects both for increased efficiencies and for fuel supply from solar biomass, photovoltaic and photochemical conversions.

It is not yet clear where vigorous policies to promote these alternatives could lead us. Certainly they could save or provide much more energy by 2000 than could a breeder economy—and at much lower unit cost. Equally certainly, they have major socio-political advantages not found in giant, high technology options such as the breeder: planning flexibility; diversity of supply; likelihood of rapid cost-cutting through innovation and mass production; low environmental impact; and a more equitable technological balance between industrial and developing countries.

Looking beyond 2000, nuclear power might well be needed to back a strategy of conservation plus renewable sources (plus coal). However, we do not *know* if this is so. Nor do we know if plutonium breeders, with all their risks, must be a major component of that nuclear power.[4] There has been little time since 1973/74, when the need was first widely recognised, to investigate thoroughly the renewable and conservation alternatives to oil, nuclear power and profligate energy waste.

These studies are now being pursued urgently. Since there is no urgency for launching into a plutonium breeder economy, delaying a commitment on CFR-1 can do no harm and might lead to much wiser, safer and surer energy options.

In the absence of Mr. Leach, Mr. Romig was questioned on his behalf:

CHAIR If we were to scrap our housing stock and start from scratch we could save a lot of domestic energy, but could you explain to the Hearings how we can find a lot of our energy needs from alternative sources and by conservation before the year 2000?

ROMIG It would be a mistake to consider that one would not use the existing housing stock. The buildings of a nation should be considered as a national resource and if they were upgraded and treated with the same sort of consideration that energy generation is given, then there would be a considerable potential for energy conservation. From insulation alone the existing dwellings stock could save somewhere between 7 and 10% of the U.K.'s primary energy—that's just from cost effective measures now. If one were to apply that to the national total and use such appliances as heat pumps and solar panels there would be a much larger potential energy conservation than is recognised at present.

CHAIR Would you explain how energy use in industry could also be cut?

ROMIG My particular area of expertise is in architecture. I have recently seen studies in the United States done by Portland Cement and various other firms, which are currently investigating the potential for energy conservation in their manufacturing processes. Some of these processes and some of their projections suggest that they could save something like 15-20% in production. Several production methods are beginning to be investigated now on a broader scale than in the past.

ADAMS Mr. Leach in his article suggests that there are safer alternatives to plutonium breeding and then goes on to give examples of the use of Thorium 232. Could you enlarge on that?

ROMIG I'm afraid I can't.

CHAIR This is important. I'm grateful to you for raising the matter. I did ask the Canadian High Commissioner if he would send someone to talk about C.A.N.D.U. but, alas, I had no reply.[5] I'm going to put this question to the A.E.A. presently since they're responsible for research over the whole of this field.

WHYTE Mr. Leach's paper refers to the major socio-political advantages in developing alternative technologies rather than dependence on fast breeder reactors. Will you tell us what these socio-political advantages are?

ROMIG I was speaking recently with some people at JPL about some of their energy work and I think that a decentralised system is best. This is much less susceptible to all sorts of societal and even strategic ills. Terrorists can attack centralised power stations but it's much more difficult to do that if they're decentralised.

CHAIR (Question from the floor). If you were given £1000 million, the estimated cost of CFR-1, to spend on back fitting current housing stock in Britain, how much energy could be saved? I think I must also add how much would be spent on doing it?

ROMIG I don't have a ready answer for that.

CHAIR We can put the question to Dr. Chapman. I think he has done that calculation. (Question from the floor). Are you acquainted with the Parker Morris standards of building and do you think they need a drastic revision in order to promote better standards of insulation than exist at present and the sort of alternative energy saving technology you advocate?

ROMIG Yes. I think the new I.H.V.E. standards would probably be adequate but they should be revised.

CHAIR I'm going to put a loaded question; do you think that the University Grants Committee should include sums for insulation in their new buildings?

ROMIG Yes.

CHAIR At least I've got that one over . . . let somebody else ask a serious question!

HENDERSON In the paper which Mr. Leach submitted the statement is made that the vigorous policies of energy conservation and the development of new energy sources could save much more energy by 2000 than a breeder economy, and at much lower unit cost. If this is true why isn't more being done already without subsidy to develop these technologies and differences in life style? Is this partly because existing institutions like power companies or our own Atomic Energy Authority have too much of a vested interest in their own techniques? Are obvious opportunities just being missed?

ROMIG It is important to take a look at the various energy needs of society and how these demands can be met in the most economical way. Then we should generate energy accordingly. This is not the same as just assuming there is a large quantity of energy required and that energy must be generated on the basis of that assumption.

COTTRELL A good part of what you're advocating comes to this. It is for an individual builder or householder to decide to spend money on insulation for his own economic benefit. Is this correct?

ROMIG Yes. There are different ways in which this can be handled. This is in the householder's best interests, and it's also in the nation's best interests. Buildings should be considered as a national resource.

COTTRELL I'm slightly confused because you've gone off the individual householder's gain to the nation's gain. Is there some difference there?

ROMIG As for the individual householder there are different schemes that could encourage tax relief. This is done in other countries. In France, for instance, if you insulate your house and put solar panels on it there is a tax relief scheme. There isn't such a scheme here. I think that perhaps the Save It Campaign has made a certain impact but perhaps a further impact could be made. There are also schemes in the United States proposed by the American Institute of Architects, whereby utilities would insulate homes, and the insulation would be paid by the customer as the energy savings accrued. The money that would normally be spent on the energy goes for the insulation, and after a while the insulation is paid for.

*Dr. Peter Chapman's submission on alternative sources was as follows:*

Our preliminary analyses of the potential supply capacities of the non-nuclear energy systems are summarised in Table II.

In addition to these supply alternatives we have also found that substantial shifts in the patterns of use of particular types of energy can reduce demand. Our study of electric vehicles shows that vehicle option has a 5-10% less primary energy demand and a 20-30% smaller electricity generating capacity. The adoption of combined heat and power schemes could also reduce total primary fuel demand by 5-10%.

We may conclude that many studies point to the large degree of uncertainty concerning future demand and supply. All of the studies undertaken at the Open University Energy Research Group show that if the F.B.R. is needed it will not be for 40-50 years. Furthermore we have found a range of attractive alternatives that may make such a hazardous development unnecessary in the U.K.

*Table II The Potential Supply Capacities of the Non-nuclear Energy Options in AD 2000 and AD 2050*

| | *2000* | | *2050* | |
|---|---|---|---|---|
| | *mtce* | *% of Demand* | *mtce* | *% of Demand* |
| Demand Estimate | 425 | | 650 | |
| Coal | 150 – 200 | 35 – 47 | 200 – 300 | 31 – 46 |
| U.K. oil and gas | 250 – 300 | 55 – 71 | 10 – 30 | 1.5 – 4 |
| Solar (heat) | 5 – 10 | 1 – 2 | 50 – 100 | 8 – 15 |
| Waves (primary electricity) | 5 – 10 | 1 – 2 | 50 – 100 | 8 – 15 |
| Geothermal (heat) | 3 – 6 | ½ – 1 | 10 – 20 | 1.5 – 3 |
| Wind (electricity) | 5 – 10 | 1 – 2 | 10 – 20 | 1.5 – 3 |
| Tidal (electricity) | 3 – 6 | ½ – 1 | 10 – 20 | 1.5 – 3 |
| Totals | 421 – 542 | 94 – 126% | 340 – 590 | 53 – 89% |

CHAPMAN If I may begin by attempting to answer Sir Alan Cottrell's question to the previous witness—it seems to me that the problem consists of trading off capital investment costs against running costs. Something like 50% of the housing stock is actually owned and built by local authorities, and local authorities in their building programmes are "capital constrained." They are not allowed to build houses that cost more than regulations permit. This means that they have to go in for the minimum level of insulation. The private market builders already have to meet legislation requirements through the building regulations in order to give the customers the necessary structural guarantees of the properties of the house. This is because the builder's own commercial interests, at least in the short term, are to build houses at the lowest capital cost regardless of the running costs. So the reason why householders and local authorities don't go in for insulation even though it is clearly cost effective is that they have not got the same access to capital as the centralised generating industries. And further, the calculation which the Chairman asked to be done about the effect of spending £1000 m. is as follows. Suppose we spend £1000 m. on a million houses, a reasonable way to spend the money, we would save 4 kilowatts on average at peak demand. The house would previously have had a peak demand of something like 10 kilowatts for heating—this is for space heating and water. We would be able to save 4 of those kilowatts. So we could save 4000 megawatts of capacity for my £1000 m. spent on house insulation.

COTTRELL What you're saying about the local authorities and builders is very interesting, does this mean that privately owned houses in general are well insulated?

CHAPMAN The access to capital by the private individual is nowhere near as good as on the part of the nationalised industries. I wouldn't like to make any general judgement about whether publicly owned houses are better or worse insulated.

CHAIR You say by the year 2000 we could meet between 94 and 126% of demand from alternatives. What kind of alternatives would be used for industry?

CHAPMAN These figures are not really based on alternatives. The largest slice by far is from coal, oil, and gas. When I specify the potential supply capacities I'm not suggesting that these are predicted supply levels. They are the potential capacities that could be realised with a concerted investment programme in all these industries.

HENDERSON Do you suggest that the fast breeder is unnecessary for satisfying reasonable projection of power demands as far ahead as one can see? You have not mentioned the notion of cost minimisation.

CHAPMAN You've inferred more from the information than was intended. I referred to the potential supply capacities of fossil and alternative fuels. Since their figures are so close to the estimated demand, and since like any other industry the fuel industry is unlikely to achieve its targets, I would suggest that we take at least one other option, say the nuclear thermal option, to back this up. On the question of cost minimisation I think that the remarks we've heard already about the cost of wave power, fast breeder reactors, thermal reactors etc. are still in dispute. There are so many uncertainties that it would be a waste of time to try to do a cost minimisation at this stage. It would seem sensible not to stretch any one source to its limit since if you do that, you are likely to incur very high marginal costs.

CHAIR (Question from the floor). Could you give an indication of the contribution that heat pumps might make to energy conservation?

CHAPMAN I find this a difficult question because it's very complex. I guess that in the questioner's mind there is the notion of an electric heat pump which, for my money, is not the best sort of energy technology. If you want to use heat pumps, I would suggest that you use an oil or gas fired heat pump. In all houses that are presently heated by electricity you could then reduce electrical demand by a factor between 2 and 3.

SEARLE You've been much more conservative or pessimistic about the contribution to be made by solar and wind power than Dr. McVeigh. Do you regard his submission as being wildly optimistic?

CHAPMAN I don't know the details of Dr. McVeigh's calculations. My own feelings on wind are that you would not be saving generating capacity by building a lot of windmills but that the only places where wind would be economic would be those where you had no alterative supply or in a remote district. That is possibly the source of our difference there.

CHAIR And solar?

CHAPMAN I'd like to make a very important point, that my calculation is not based upon flat plate collectors. It's based largely on sensible house design and so called "passive solar technology" where you have large south facing windows to collect the sun's energy. Today we already receive something like 10% of the heating demand in most houses through incidental solar gains (that's the name given to them). If one is a little more careful architecturally in the design of the orientation of the house, say, if you have your living room on the south side and rooms like the bathroom and kitchen on the north side, this proportion of 10% can be quite easily increased to 20 or 30% at zero production costs.

CHAIR Dr. McVeigh, is there anything you wish to say in reply in just one minute?

McVEIGH The economics of wind power are grossly distorted

in everybody's calculations because they still will not accept the concept of having a very large number of cheap mass-produced windmills on the very large area of the country that has not very good wind sites. These could make a very big contribution to our national energy demand.
CHAIR You ought to go on that programme "Just a Minute"! Dr. Chapman, would you suggest that we go ahead with CFR-1 or not?
CHAPMAN I would suggest that at the present time there are many ways in which that £1000 m. could be spent to more profit both to the industries that would construct it and for the people that would receive the benefits from the output.

*The U.K.A.E.A. submission on alternatives was as follows:*

A decision to build CFR-1 is simply a commitment to gain experience of the problems of building and operating a full-sized power reactor of the breeder type. The decision in no way makes the country less able to meet energy needs up to the year 2000. This is because:

(i) Over half the cost involved can be attributed to building and operating a power station and some of the remainder to investment in a fuel production facility. The output from this station substitutes for electricity which, at the margin, would come from stations fuelled by oil that would either have to be imported or which could have been exported.

(ii) The balance of the expenditure represents investment in the further development of an energy technology that is very close to exploitation and which offers a cheap, clean and safe substitute for existing power station fuels that will—in its absence—become very scarce and costly.

The conclusions reached in recent studies of the feasibility and potential of many of these sources may be summarised as follows:

(a) Solar and geothermal sources are potentially capable of being used by 2000 to satisfy part of the demand for low grade heat, essentially hot water for heating purposes, but are very unlikely to produce electrical power economically. Solar heating would be seasonal.
  (i) Solar heating will probably not provide more than 5 m.t.c.e. (million tons of coal equivalent) by 2000.
  (ii) Geothermal: the extent of this resource will not be known until geological explorations are complete in about 3 years.

(b) Tides: In some configurations tidal power can be combined with a storage capacity to provide a firm power supply. The only large site available in the U.K. is the Severn Estuary, which would pose formidable and unique engineering problems which have not yet been assessed in detail. Different schemes suggested would give fuel savings between 3 and 10 m.t.c.e.

(c) The sources of wind and waves can be considered together as potential power producers. They both have the same characteristics of a variable and unpredictable power source which, although at their largest in the winter, cannot be relied upon to meet the pattern of demand placed upon them by consumers. It follows that their use might save fuel but not the building of power stations. Their capital costs and maintenance charges must therefore be offset against a potential saving in fuel charges only. It is unlikely that it will ever be economic to use them instead of the same electricity from base load nuclear plant but it may prove economic to conserve scarce fossil fuels in this way though stand-by plant will have to be built for use at dead periods.

(i) Wind. The annual average output of generators on 10,000 sites might save 6 m.t.c.e.

(ii) Waves: Waves on the north and west coasts constitute a large resource. It might be possible to generate electrical eneregy up to half the present consumption if devices could be built along a 600-mile stretch of coast, perhaps 10 miles out to sea (and bearing in mind the economic penalty of installing an equal amount of stand-by-plant). This would be a formidable engineering development and the potential supply by 2000 may lie in the range 5-15 m.t.c.e.

The energy sources listed above are at an early stage of development and there is a lack of firm economic data in each case. It is far too optimistic to assume large-scale use by the year 2000. They could not alone supplement the fossil fuels likely to be available at or after 2000 sufficiently to meet the expected demand for energy.

None of these options is affected by the decision to build CFR-1. They are all at a very early experimental stage in their development when only relatively small sums of money can be absorbed to useful effect.

Dr. L. E. Roberts, Director of the Atomic Energy Research Establishment at Harwell, was questioned as follows:

CHAIR Am I right in saying that you're not so optimistic as Dr. Chapman and Dr. McVeigh on the subject of alternatives?

ROBERTS That's true. Our guesses about the contribution that various alternative sources might make by the year 2000 were put together by the Energy Technology Support Unit in Harwell after consultation with other bodies as well. The subject should be further researched. Many of these sources are now at a quite preliminary stage of their development. Many of them—I'm talking about solar, geothermal, wind and wave energy—can be developed further, up to the point at which the unit of power operating would be full-sized without a very great commitment in resources because the unit size would be small. The unit size of Dr. McVeigh's wind generators might be 1 megawatt, the unit size of solar panels a few square metres, the unit size of Mr. Salter's wave devices generate about 10 megawatts perhaps at most. And if we have any geothermal heat worth utilising—as you said, this isn't certain yet—the unit size of a heat mine might also be a few megawatts thermal output. All these are small unit sizes and all of these sources could be pursued by this country in conjunction with other countries who are also investing in these technologies, up to a really full-sized demonstration. From this one could gauge the engineering feasibility and the economics with some precision. The development of these resources will not be precluded by a decision to go ahead and develop CFR-1 since they are of a different order of magnitude.

ADAMS How much research on these has been done by the Atomic Energy Authority at Harwell?

ROBERTS At the moment we're spending perhaps three-quarters of a million pounds per annum in support of the Department of Energy. I think the Department is spending about two million this year and taking part, of course, in international programmes as well. Therefore we will have the benefit of much larger resources also from other countries of the E.E.C. and also in America. Other government departments are also supporting work on alternative sources of energy.

CHAIR We were interested to hear you say that the Secretary of State was disbursing two million pounds. Could you help us to know, roughly, how that compares with the amount of money being spent on nuclear fission?

ROBERTS I think that the amount spent by the Authority on R and D last year was £90 m. (and will be about £114 m. this year).

CHAIR Do you feel happy about that proportion?

ROBERTS It is my contention that the alternative sources (excepting fusion) can all be developed to a stage where we can take sensible engineering and economic decisions about further investment in any of them for comparatively small cost because the unit of development is small. I therefore think that the ratio is not significant.The late development costs in atomic energy and nuclear power up to, perhaps, a decision to invest in CFR-1 mark a final stage of development in a technology which is very highly developed, and where—if we built CFR-1—we would be gaining a modern generating station with an output of 1300 megawatts. That scale of decision, which is appropriate at a late stage of a technology, is not appropriate for any of these others at this time. When they have proved themselves in large scale tests, perhaps ten years from now—even earlier, if Mr. Salter is correct—that would be the time to decide when massive investment is required.

SEARLE We heard from Dr. Chapman that if he had £1000 m., he would spend it on a million houses at £1000 each, and reduce from ten megawatts to four megawatts the energy demand. This would be a nett saving of six thousand megawatts. For the same amount of money you're producing 1300 megawatts. This would be development, not research. It would be the use of a less sophisticated technology. If you had to choose between the two, which would you choose?

ROBERTS One must look at energy requirements as against energy demand. The pattern of energy demand is, of course, complex. Nuclear power and CFR-1 have been designed to satisfy a base-load demand for electrical power. This is, of course, vital to keep industrial life going, and it is also of use domestically in this country. Sources of variable supply can complement such a source, but not displace it. As for other investment aimed at the more efficient use of energy, everyone must be in favour of it. It is a matter of individual decision in the light of the benefits that individuals and individual companies might gain by this sort of investment. Benefits to be gained by insulation or any other means should be widely researched and disseminated so that those decisions can be sensibly taken.

CHAIR We are going to ask your colleague Mr. R. L. Nicholson, to answer a question from the floor on the thorium breeding cycle. (Question from the floor). Why does a thorium breeding cycle (avoiding plutonium) have to involve C.A.N.D.U? Why not thorium in the fast breeder reactor?

NICHOLSON There is no particular reason why not. You can use a thorium U233 cycle in a fast reactor. There are still problems yet to be investigated and worked out on the use of the thorium U233 cycle. The U233 is similar to plutonium as a bond material. It's more radio-active than plutonium, though less of an inhalation hazard. It poses considerable problems on the reprocessing of thorium fuels and the making-up of uranium 233 into new fuel elements to put back into reactors to make more power. It would therefore take quite a lot of development to get our knowledge and expertise on the thorium 233 cycle to the same stage as that of the uranium-plutonium cycle.

COTTRELL What do you think about C.A.N.D.U. as a thorium breeder reactor?

NICHOLSON It has potentialities. A number of thermal reactors can use thorium for breeding. Most of them need a lot of Uranium 235 as well. C.A.N.D.U. needs less than almost any other, but it has a very big inventory in the reactors, and you'd have to re-cycle; and to do this you'd need to solve the problems I've just mentioned. May I clarify a further point? Dr. Roberts gave the total R and D spent by U.K.A.E.A. on nuclear energy. I believe that the comparison may have been with fission reactors. The amount spent on fission was about £80 m. out of £114 m.

CHAIR Dr. McVeigh, do you wish to come back?

McVEIGH About the point on wind being a fluctuating source of supply; we need more work on energy storage, and then large-scale, high temperature, high grade energy storage could iron out these fluctuations. It would make a very considerable impact.

**International Aspects of CFR-1**

*Mr. Brian Johnson's submission was as follows:*

A U.K. decision to proceed now with the development of CFR-1 could seriously damage international prospects for a moratorium on the spread of "weapons-sensitive" fuel reprocessing facilities.

The initiative for a pause to consider the implications of an international plutonium economy—and alternatives which could avoid the greatly increased hazards of nuclear weapons proliferation which this must imply—came from the United States in 1975: an initiative that was strengthened and clarified by President Ford in a policy statement of 28th October, and which we may expect to see at least matched by Mr. Carter.

Together with her European partner members of the Nuclear Suppliers Group (N.S.G.), Britain has generally sought to tone down the American commitment to an effective moratorium on exports of reprocessing and enrichment technology. Nor does Britain share the American aim of getting acceptance by nuclear importing states of guaranteed supplies of "non-sensitive" low enriched uranium, so as to hold nuclear electricity to thermal reactors which do not require highly-enriched uranium or plutonium—the weapons-grade materials which, on the basis of their 1975 plans, about 20 countries aim to acquire through their own fuel reprocessing by the mid-1980s.

Britain's present policy in the N.S.G., as described by Mr. Callaghan in a Written Answer (Hansard, 31st March 1976) is non-committal as regards our export of nuclear equipment, material and technology in general. The Prime Minister would only say that we would "study each case on its merits" while relying on I.A.E.A. (or I.A.E.A. verified) safeguards. As regards British export of "sensitive" equipment and technology (which could lead to the construction of uranium enrichment plants, reprocessing plants or heavy water production plants), Mr. Callaghan would only say that "in general we shall exercise restraint", while adding that Britain is contributing to the I.A.E.A.'s study of the feasibility of including such plants in regional fuel cycle centres in the future.

None of Mr. Callaghan's assurances excluded the possibility of Britain exporting commercial fast reactor technology. Yet it is possession of fast breeder reactors which best justifies nuclear, or proto-nuclear states like Brazil, Pakistan, Iran or South Korea, in acquiring the bombs-potential technology of fuel reprocessing.

At present the United States, Canada and the U.S.S.R. (and possibly some other concerned members of the N.S.G.) are reported to be pressing France and Germany to revise their agreements to sell fuel enrichment and reprocessing plants to Pakistan, Brazil and South Africa. A British decision to proceed with CFR-1 would strengthen the hands of the French and German nuclear industry who are the proponents of selling full fuel cycle facilities to these countries, none of which has signed or ratified the Non-Proliferation Treaty.

Moreover, in view of Britain's economic situation and the unit cost benefits which will be seen to accrue from overseas sales, it seems clear that a decision to build CFR-1 would rapidly put Britain into the Market as a prospective exporter of fast breeder reactors. Apart from weapons proliferation dangers that must attend such a development, the British government should also consider the appropriateness of such an immensely costly, complex, centralised, uncertain and potentially risky technology for countries where widespread rural poverty is at the core of the development problem: countries whose balanced development will be damaged rather than helped by such urban-industry-focussed and capital and foreign exchange-hungry sources of power.

Finally, the sale of fast breeder reactors abroad, which is a possible consequence of starting on CFR-1, could adversely affect British plans to offer our own reprocessing facilities to other countries as a non-proliferation measure. For the spread of breeder reactors can only increase the "weapons option motive" of developing countries, which will increasingly want to have control over the full fuel cycle, rather than ship their fuel for reprocessing at Windscale or elsewhere: a policy that is sometimes justified now on safeguarding grounds in an international regime of "thermal-only".

Mr. Johnson answered questions:

McKNIGHT In your recent contacts with the International Atomic Energy Agency, what opinion have you formed of the efficiency of their safeguard system in preventing the diversion of material for military purposes?

JOHNSON I was shattered in reading the evidence provided to the United States Congress and the recent concern in the United States about the problem of diversion of fissionable materials, particularly plutonium. When I looked into it in a recent study I did with the International Atomic Energy Authority and with Euratom I found nothing to reassure me at all.

RICHARDSON The purposes of this meeting is to discuss whether Britain should go ahead with CFR-1. If Britain did not how would that stop other countries doing all the things they're going to do anyhow?

JOHNSON Over the last year, at the joint instigation of Britain and the United States, a grouping of nuclear exporting countries was formed, which was originally called The London Club and is now called the Nuclear Suppliers Group. It's grown to 15 members, and it's apparently beginning to have some tentative meetings with potential future nuclear suppliers; that is to say, Brazil specifically, also Pakistan and Iran who've been invited to meetings. Nuclear proliferation was a major campaign issue in the United States Presidential election this year, and a critical one in the Swedish election. The United States feels very concerned that there should be a pause in the whole process of adopting the "plutonium economy". There is a chance of this succeeding. A number of countries are working on breeder reactors, but there is no commercially operating breeder reactor at the moment; and although France and West Germany have signed agreements with three countries which are not members of the Non-Proliferation Treaty, Brazil, Pakistan and South Africa, there is a possibility—I understand from members of the British Foreign Office, a real possibility—that they can be persuaded to stop this particular set of agreements if alternative arrangements were made.

ADAMS In view of the concern which apparently exists in the United States, is there any evidence that America has cut back her programme on fast breeder reactors?

JOHNSON Yes. They've halted it.

McKNIGHT If the advanced countries proceed with building prototype fast reactors of commercial size, each country with a thermal reactor will want its own reprocessing plant to save the plutonium for the day they get a fast breeder. Can you envisage any international institution which might take over the ownership and operation of all reprocessing plants? I'm thinking of the International Telecommunication Space Organization as a parallel.

JOHNSON It was a proposal put up by Dr. Kissinger at the United Nations in September 1975, when he called for studies by the I.A.E.A. and other governments of how regional fuel cycle facilities—particularly the weapon-sensitive ones, the reprocessing plants—could be organized. The British Government is very interested in this, with Britain strongly supporting these studies. The British Government might put forward the idea that Windscale should become a regional reprocessing facility for part of the world, depending on how much re-processing is offered by the United States or other countries. There are tremendous problems here. One of these is that if you go ahead and try and set up, say, a Latin American regional reprocessing facility by a multi-lateral agreement, you may in fact be hastening countries towards a situation where they are getting into the "plutonium economy". There are several major problems in getting such an agreement organized. Our experiences with multi-lateral projects under Euratom and under the N.E.A.—what used to be the European Nuclear Energy Agency—has been dismal. There is also the problem that any country that's interested in this, and will put up the money to back it, for example Brazil, is a sovereign state, and if it really decides to go ahead, it can just take the whole thing over and throw everybody out. You really don't have very full safeguards.

WHYTE The argument was put forward yesterday that one of the advantages of proceeding with CFR-1 and Commercial Fast Reactors is that we would thereby be leaving more other sources of energy for the developing countries. We would be indirectly benefitting these developing nations. Are you saying that we would really be doing them harm because they would be tempted to imitate and adopt an entirely inappropriate technology for their own needs?

JOHNSON There's a great deal said about the powers of imitation in developing countries. I prefer the term 'ventriloquism', because what happens is that the International Atomic Energy Authority and a number of governments have been going around the world creating nuclear bureaucries or military bureaucracies in developing countries. These naturally club together with the I.A.E.A. and decide that they would like a nuclear industry. Probably they won't get it in most cases because they can't afford the foreign exchange, although there are 20 countries now in the developing world alone which claim that by the late 1980's they will probably have fuel reprocessing facilities, which will give them nuclear bombs if they want them. I'm alarmed by that kind of picture of terrifying proliferation because the history of the nuclear industry generally, is "promises, promises . . ." We were told in the late 1950's, for example, that 25% of our nuclear electricity would be produced by fast breeder reactors by the late 1970's.

WHYTE I was thinking of the development that's required in countries whose problem is rural poverty. This is a different kind of development that they're really requiring. I was referring to peaceful rather then military use.

JOHNSON One of the critical factors is the appropriateness of the technology. There are much more appropriate technologies for developing countries then those we have discussed as alternative technologies. We're going gradually to find that, as we reach the real barriers, which haven't even been discussed here—the heat barrier of the 21st and 22nd century, which the International Institute for Advanced Systems and Analysis is now doing studies on—that we just cannot go on generating electrical energy.

CHAIR I think we must keep to the 20th century.

JOHNSON Sometimes you have to take the very long view to see whether you're on the right path.

CHAIR Will you please explain why the sale of fast breeder reactors abroad would adversely affect the plans to offer reprocessing facilities at Windscale to other countries.

JOHNSON The problem is that developing countries would

want to have the full fuel cycle so as to get total independence for strategic considerations. Britain has hoped to be able to offer to reprocess their fuel, so as to keep them out of that world. There is an incompatibility of two logics here, but I haven't seen it argued publicly.

*The submission of Mr. Maurice Garner, formerly Undersecretary of the Electrical Division of the Department of Trade and Industry, was as follows:*

A hard-headed look should be taken at all proposals for participation in a European combined project.

I do not overlook the problems all this implies for the British nuclear engineering and heavy electrical plant industries. For several years I have regarded these problems as insoluble outside of a programme of collaboration involving the Government, the electricity boards and the plant manufacturers, and involving for the manufacturers concentration, modernisation and merger or association with interests in Europe or the United States—for preference, the latter as, given the costs of development of nuclear technology, I doubt whether such development will ever be economic except on the basis of a world market and this I think the Americans more likely to secure than the Europeans, if only because the Americans seem to be freer with public money.

The current soliciting of American partnership in future aircraft development by both Britain and Framce separately reveals a parallel with the nuclear engineering situation and shows the way the wind is blowing.

It is to be hoped that the C.P.R.S. who are studying the problems of the heavy plant industry will not do so simply on the national plane but will come forward with proposals for dealing with our national difficulties that take account of the forces operating now on the continental, if not world plane.

The message of this essay is quantification, comparison of costs and benefits, systematic definition of all the available options (including manipulation of future electricity demand by using the price mechanism and other energy policies) and proper allowance for risks and uncertainties.

Mr. Garner spoke as follows:

My concern in these matters is to get the best return for the public on public money, and this involves the systematic examination of all available options and the application of common standards to competing programmes of public expenditure. Earlier I put forward what could be regarded as a *prima facie* case suggesting that CFR-1 was not the most advantageous application of public money, but I am aware that there are some factors that are not so readily quantified, for example, the convenience of the absorption of foreign technology and the export potential. As for the first point, when the issue of nuclear choice was last before the country—over the light water reactor or the S.G.H.W.R.—neither of the two bodies most affected, the Central Electricity Generating Board or the National Nuclear Corporation, seemed to see any practical difficulties about the absorption at that stage of foreign technology. Secondly, as for export potential, it's important to find out precisely what the benefit to the country will be from the export of our CFR-1 technology if it becomes available.

SEARLE It's been argued by Mr. Johnson that nuclear technology is inappropriate to the potential market of the developing countries. Would you agree with him, or would you think we ought to take a moral and ethical stand as well as just an economic one?

GARNER I certainly think one should take a moral and ethical stand as well as an economic one. I was merely concerned with the economic argument. When I refer to a world market I ought to make it plain that I'm not talking about a world market for just British manufacturers. In my paper I referred by analogy to the aircraft industry as I believe this shows the magnitude of the problem that is now confronting us. In an essay which appeared in an American book called *The Impact of Large Firms in the U.S. Economy,* two Americans wrote "widely available data suggests that for each new generation of modern aeroplanes being produced today, it is doubtful whether the entire world economy can support the profitable operations of more than one, and certainly not more than two producers of a given family of airborne transportation systems." This was where Rolls Royce fell down: by the time they were producing their advanced engine they were up against the situation that the world could hardly sustain two aircraft engine manufacturers. I am asking people to face up to the possibility that we may be moving into the situation where, in economic terms, there is a prospect of getting a genuine economic return on a nuclear system only if there is one nuclear system pretty well for the whole world.

CHAIR But are you content with the idea of exporting fast breeder reactors outside this country to other nations?

GARNER I don't think that I am an expert witness on that subject. That seems to me to be a moral question and a question of environmental safety.

CHAIR Do you think that the moral question can be divorced from the economic one?

GARNER Yes, it can be divorced in the sense that I think it is right, first of all, to ask: Is what we're proposing to do with public money an economic proposition? Only when we're established that it is an economic proposition do the moral questions arise as to whether it's desirable to do it.

COTTRELL We have been told by the A.E.A. that they envisage a programme of fast reactors of 33 Gigawatts eventually, which will grow out of a CFR-1 of 1.3 Gigawatts, plus what's gone before with the P.F.R. Is that not, in economic terms, a satisfactory ratio of production to research and development?

GARNER The answer is No, especially when spread over the time scale involved. I would refer you to an earlier article I wrote in the *Electrical Review* two and a half years ago, in which I demonstrated then—at least to my satisfaction, and nobody challenged it—that for the reasons which I have given, nuclear development, certainly in this country on an isolated basis, cannot be economic because of the enormous cost of the research, and the long time scale, and the period over which the ultimate construction of the stations has to be spread.

ADAMS Do you consider yourself on expert on the issue of importing other people's CFR's?

GARNER In economic terms, yes. I've said 'yes' to imports and 'no' to exports in purely economic terms.

RICHARDSON Although the world market may not support five conflicting systems, there seems a lot of pressure within the industry for the West European countries, which are a long way ahead of the United States, clubbing together and forming a co-operative venture. Are you dismissing this and suggesting that we should pull out of the business altogether?

GARNER I'm not dismissing it. I'm merely saying that one should look in an unblinking way at the economics of a European scheme as much as one should look into the case for a British scheme. I have doubts because of the enormous expenditure over a protracted time scale even in a European scheme. But I can't prove it, and I would welcome studies being undertaken in order to assess the economics in a proper way, using total resources and calculating the ultimate benefits, and discounting the benefits against the expenditures.

Mr. Surrey and Dr. Dombey replied to further questions, this time with particular reference to the economic aspects of CFR-1:

COTTRELL Do you regard a ratio of 33 Gigawatts of eventual production to 1.3 Gigawatts from CFR-1 as unfavourable?

SURREY I would like to answer that in exactly the same way as Maurice Garner did. The question concerns the time taken to reach 33 Gigawatts, and the rate of discount that's used for it. The question is whether you take the 10% rate of discount which is used by the Treasury for all public sector investment or whether you take some other rate of investment, and what the justification for taking a lower rate of investment is.

ADAMS Do you believe that licensing from abroad might be the best method if supplies of uranium become very short?

SURREY If uranium does become very scarce, then if fast reactors are developed anywhere in the world and introduced in substantial quantities, this will help to provide an upper ceiling on the level to which uranium costs can escalate. CFR's don't necessarily have to be made here. If the Americans make them in quantity, then that alone will provide insurance against the kind of risk to which you're referring.

HENDERSON Have you anything to say about the possibility of collaborative arrangements?

DOMBEY This would largely be a matter of bargaining and therefore I don't have very much to say about it. But we do know that the history of international collaborative projects hasn't been happy. It's really a question of the terms which you can negotiate. The energy problem of the French is much more immediate than here, and they are committed to going ahead with Super Phoenix. It would make sense for Britain to try (a) to use the P.F.R. as a tool to learn as much about fast reactor systems as possible and (b) to negotiate with the French a modest share in Super Phoenix. The French need a large amount of plutonium for Super Phoenix, which we may be able to provide for them as a bargaining tool.

*The submission by the National Nuclear Corporation and Nuclear Power Company Ltd. was as follows:*

Breeder reactors are being developed in Russia, Japan, France with Italy, Germany with Benelux, and the U.S.A. It does not therefore seem likely that a decision for or against the construction of a U.K. demonstration reactor (CFR-1) will of itself have consequences which influence the adoption of fast reactors in these countries or subsequently in ohters. In any case, N.N.C./N.P.C. visualise the demonstration programme for the fast breeder as a concerted Western European, if not a Western world, objective. This would be a contribution towards Western energy self-sufficiency, an incidental effect of which should be the liberation of some supplies of fossil fuel resources for countries in which, for one reason or another, nuclear stations are unsuitable.

The West might also wish to export fact reactors. Such a course will obviously be a decision of Governments: it may well be that the U.K. will best influence such decisions if it has participated in the development and demonstration programme of the Western community.

On the question of spent fuel reprocessing requirements, a four-part answer is required.

(i) Nuclear fuel is of various sizes and chemical composition: the most appropriate single index for measuring the "amount" of spent fuel is the contained fission products. The quantity of such fission products which have to be separated from irradiated fuel is closely related to the total heat production from a reactor, whether thermal or fast. Our answer is framed using fission product quantity as the index.

(ii) The heat output from a first U.K. demonstration reactor (CFR-1) is likely to be around 3,000 MW (Thermal). The corresponding figure for the pre-existing U.K. reactor programme is likely by that time to be about 36,000 MW (Thermal). Accordingly, the demonstration reactor will increase the rate of arising of fission products in the U.K. by about one-twelfth.

(iii) In so far as CFR-1 establishes an option for a fast reactor pro gramme in the U.K. and that option is taken up, there will be no additional production of fission products if the fast reactors are constructed instead of thermal reactors.

(iv) In the longer term the production of fission products (and electrical power) from thermal reactors is limited by the economic availability of uranium. If fast reactors are adopted it will be because they offer a virtually inexhaustible supply of fuel independent of other parts of the world; to the extent that this supply is utilised, fission products will be produced. Of course, production would take place over many years, during which time most of the earlier produced fission products would have decayed to non-radioactive forms.

Dr. Franklin was cross-examined as follows:

CHAIR Would you like to comment on any of the observations which have been made?

FRANKLIN There are a number of important points. First I too am very concerned about proliferation. I believe that the question of proliferation is only partly related to nuclear power. There is sufficient information published on reprocessing operations, particularly simple reprocessing operations, and simple kinds of reactors. I therefore see no reason to believe that the proliferation problem is going to be determined by the will, or otherwise, of a suppliers' club, to exclude other countries from reprocessing nuclear fuel. Further, the technological problems of reprocessing fuel which has been long irradiated in reactors, though not outstandingly difficult, are nevertheless not those you would choose to engage in if you were trying to make illicit weapons. There has been a tendency to suppose that if the advanced countries tried somehow to impose their will upon other countries who felt that they wanted—for one reason or another—to have a supply of plutonium, then they would be able to do so. I don't believe that this is true. There must be half a do zen people in this hall you could hire, and you could have a decent reprocessing plant within five years. On the question of recovery of funds as a result of investment in R and D, the last speaker made the point that the true test discount rate was all important, and indeed it is. And of course if you do sums on the basis of 10% or 15% and you say that's what you want, without taking account of inflation, then almost any high technological work at present—unless the market is enormous—will flounder. It will even flounder if you have one single world market available to you. But indeed the test discount rates which are used have implicit in them assumptions about inflation for the future. And you ought to add that to the analysis as well, and then you would come out with some different sorts of results.

On the question of licensing, I think it is very difficult to know whether a licensing avenue is the best one. Certainly so far as the National Nuclear Corporation is concerned, we have not for years advocated a posture in which the U.K. went ahead alone with a fast reactor. We have regarded the likely outcome for the world—to quote a previous speaker—as one in which perhaps two nuclear companies could supply the fast reactors to whatever was considered by Western statesmen to be an appropriate market; and I have in mind the advanced countries, and not the world in general. The question then as to what these organisations might be and what the British position might be in them, is one in which I have personally been involved over the last few years in negotiating with the French, Germans and Americans. I can only say that we ought to be able to negotiate an agreement, and that in my view in due course it will be necessary to build a demonstration fast reactor in the United Kingdom, because without going through the exercise—the licensing process, the siting process, and the other things which are involved—you will not have the option because you will not know that such an operation is acceptable. On a question of fact concerning Uranium 233 breeding, I refer to the suggestion that Uranium 233 be diluted with 238 in order to render it virtually useless as a bomb material. If you then put that mixture back into a reactor you will inevitably generate plutonium from it.

SEARLE In your paper you say that an incidental concomitant of development of the fast breeder in the U.K. would be—or should be—the liberation of some supplies of fossil fuel resources for countries in which, for one reason or another, nuclear stations are unsuitable. Which countries are they unsuitable for? Why? Do you really think that this development will result in the liberation of fossil fuels for the less-developed countries?

FRANKLIN The question of unsuitability can be of two kinds. First it is very likely that fast reactors are going to be relatively large unit stations. If you take account of the present position or even the prospects for a great deal of the world's population it is most unlikely that by the year 2000 or thereafter they're going to be equipped with a grid system which is capable of accepting a unit size of this sort. Therefore on purely technological grounds I think fast reactors might be unsuitable. But I must say this. If there were a question of exporting fast reactors, such a course will obviously be a decision of Governments. I think that Governments might also come to the conclusion, in the western community, that there were places to which they did not wish to have fast reactors exported.

SEARLE What about fossil fuels?

FRANKLIN If you look at the American situation in relation to hydrocarbons, and after all, American consumption absolutely dominates the world scene, the position of O.P.E.C. has been reinforced by the failure of the American attempts to economise in hydrocarbon fuels. In fact American consumption of petroleum is now about 50% imported, and the absolute quantity of petroleum produced indigenously in the States is going down. If one sees that continuing, the position of the sellers will be strongly reinforced in the petroleum market. If therefore the Americans, above all, are able to bring fast breeders into operation, it will reduce their demand, which is the decisive demand. It may also improve the bargaining position of purchasers of petroleum in poor countries.

McKNIGHT The international organisation for the gas centrifuge seems to me to be working pretty well. What worries me about the fast reactor is that your European collaboration will come at a much later stage in development than happened in that successful example. Would you care to comment?

FRANKLIN Yes. It is true it is easier to make collaborative arrangements at earlier stages of research and development, and later it becomes more difficult. But in the past it has tended to become more difficult because you had got national commercial interests strongly entrenched which couldn't find a way of reconciling their positions. The situation on the fast reactor is that in the countries where it's being developed it is, to a large extent, an instrument or a product of Government. If there is a government will to collaborate, then means can be found industrially.

RICHARDSON The Energy Secretary was at pains to say at these Hearings that there was plenty of time still before a CFR decision and that time was on our side, and that we wouldn't lose anything in the race for international co-operation because he'd heard many times in the past that we were running out of time and this was never true. In other words, he was relaxed about the time scale. Are you equally relaxed?

FRANKLIN That's a difficult question to answer. I'm obviously not relaxed in one sense because I'm in charge of a group of people with know-how in this area, and I'm afraid that that know-how would be dissipated if they were left without a task for too long. But it has been difficult for me to ascertain whether the Secretary for State, when he spoke in these terms, was referring to a few months or a few years. If it's a few months, then quite certainly the matter is sufficiently important that it can wait that length of time.

CHAIR It wasn't Mr. Benn, was it, who said that a week is a long time in politics? I think it was somebody else!

FRANKLIN If there is going to be a collaborative arrangement for the development of the fast reactor, it will have to be arrived at by negotiation. There is no point in going into it and saying, "We're going to start our fast reactor on a particular date," because unless you're first, the date must take some account of the findings of what has gone before.

CHAIR (Question from the floor). On the 29th October 1976, President Gerald Ford called for international co-operation to halt for at least three years the re-cycling of waste. Is there any indication that this has been taken up in this country?

FRANKLIN What he actually called for was termination of the recycling of plutonium since in the normal way one doesn't recycle nuclear waste. As to the intention of this country, it obviously depends a great deal on overseas reprocessing contracts.

CHAIR Thank you very much; would Mr. Johnson like to reply to what you have said on the subject of proliferation?

JOHNSON Of course we all share this concern for proliferation. I didn't stress adequately that if you get into the business of commercial fast reactor production, if you are to lower your unit cost—as with selling aircraft or weapons overseas—you have great pressure to export. We are now at a delicate stage of negotiations with other countries whether or not other countries go ahead with their plans to sell reprocessing facilities and fast reactors. Therefore if a decision were made shortly in this country it would tend to strengthen the hands of other people who are doing things that we might not want to do internationally ourselves. The general problem of fissionable materials lying around the world is very worrying. There's "the nuclear overhang", which describes countries which don't have reprocessing facilities, but if they got them—if they had backyard ones as was suggested, or made them quickly—they would then have large weapons capabilities. I would suggest that this is not a very cheerful prospect for the nuclear industry at all, but that is another question.

CHAIR Dr. Franklin, do you feel you've got anything to say in reply?

FRANKLIN I can only repeat that the fast reactor is not in any significant way associated with the problem of proliferation.

*The U.K.A.E.A.'s submission on international aspects of CFR-1 was as follows:*

While a decision by the U.K. to proceed with CFR-1 would allow a significant addition to world experience in the operation of commercial-scale fast reactors and their fuel cycle, we should not expect other countries to choose to adopt or abandon fast reactors as a simple consequence of our policy.

As far as reprocessing is concerned, the operation of CFR-1 itself will increase the total amount by a few per cent since it will initially supplement the generating boards' power station capacity and will not immediately displace a thermal reactor. However an ongoing fast reactor programme, in which each new fast reactor substituted for a new thermal reactor, would reduce the amount of reprocessing required for a given electrical output since each tonne of fast reactor fuel generates much more electricity than a tonne of thermal reactor fuel. (While fast reactors do introduce new technical problems in reprocessing as result of the higher plutonium content of their fuel, by the time CFR-1 comes on stream these should be well understood, since by then the U.K. will have had close on ten years experience of reprocessing P.F.R. fuel at Dounreay).

CHAIR Mr. Nicholson, would you like to add anything further to Dr. Franklin's point?

NICHOLSON On international aspects, I would confirm the view expressed in our submission and echoed in different ways by Dr. Franklin. If we were to stop or long postpone the CFR-1, I have some doubts whether that would make a significant difference to the fast reactor development in other countries. I think Mr. Johnson said that fast reactor development had stopped in the United States. That is not correct. The Ford announcement was that they would take a decision on their commercial fast reactor programmes—that is, installing them for power generation—not before 1986. The Clinch River breeder reactor project and their other fast reactor work, which is a very large programme, is carrying on just as before.

ADAMS Are you referring to the statement by President Ford, "My decisions today do not affect the U.S. programme of research and development on the breeder reactor?" That programme assumes that no decision on the commercial operation of the breeder reactors which require plutonium fuel will be made before 1986.

NICHOLSON Yes.

CHAIR Mr. Johnson may feel that he's been misrepresented. Will he reply?

JOHNSON It is true that the President said that in his statement of the 28th October, but the current budget with funds for the American fast reactor research programme is held up in Congress, where it has not been passed, and the reason for its being held up is concern over that part of the programme. So it is true to say on the best information I have, that the future of further development of the commercial fast reactor in the United States is heavily in doubt at the moment.

**Investment—CFR-1 and Alternatives**

CHAIR I think we should turn to the crunch question which concerns the effect of our commitment to CFR-1 on alternatives. But I want to put two questions to Dr. Chapman. First of all I want to ask him what, if any, are the economic implications for alternative energy sources of developing CFR-1?

CHAPMAN In the very long term this country has only a certain amount of money available for investment in projects of all types. The electricity industry has historically taken ten to fifteen per cent of gross fixed capital formation. If anything of the order of the A.E.A.'s programme for the development of reactors were to take place, this would absorb—under most G.O.P. projections—something between twenty-five and thirty-five per cent of all industrial fixed capital formation. This means that they would effectively be starving funds, not only for alternative energy sources, but also for industry. That's on very long-term availability of investment funds.

CHAIR The second question is what are the implications for research and development funding if CFR-1 is agreed?

CHAPMAN This refers to the short term problem. It is a question whether R and D funds which are spent on one project could in fact be transferred to another budget. In practice from the way that the Government—as I understand it—allocates R and D budgets, it is unreasonable to presume that closing down one R and D project would, in fact, allow another one to start up. I suppose, at one extreme, if one envisaged the U.K.A.E.A. being wound up as an organisation altogether, that wouldn't mean that you'd immediately have lots of money to spend on windmills. You might have lower taxes, more hospitals, or just more Civil Servants. So although the interchange of funds between one R and D budget and another is possible, I don't think that this is a real issue. What might be a real issue is that Ministers may well perceive that if they've made a massive investment in something like CFR-1, then they have done enough to satisfy our future energy needs. It's a matter of practical politics rather than economic theory.

HENDERSON The normal test of an investment like the CFR-1 is whether it will pay off at a given rate of discount. If it would pay off, when the CFR and later breeding stations came in, would not the substantial fuel savings which this would make possible give us greater national resources from which we could do more investment than would otherwise be possible?

CHAPMAN The amount of investment that you are able to do, regardless of the return you will get on any one investment, must be limited by the availability of capital funds, which is related to people's free and forced savings. And even if I could present you with a million projects, all of which would show a 15% rate of return, you would not have the capital availability to invest in all of them. You'd have to make a choice. I'm saying that on the basis of something of the order of 20% of G.O.P. being available each year for investment, if you try to build up a very capital-intensive programme you may—even if you meet your test discount rate—starve other projects which would meet the same test.

CHAIR Dr. Roberts, would you like to speak here for U.K.A.E.A.? You are in favour of committing large sums of money to the CFR-1. We would like to know your views on whether you feel that this will inhibit investment in other parts of the economy and, in particular, in alternatives. If you think it will inhibit it, do you think it's justified in so doing?

ROBERTS I personally, and the A.E.A. of course, are in favour of a commitment to proceed to build CFR-1, on the grounds partly described here in this submission, and partly in our earlier ones. Here I make the point that in deciding to build CFR-1 you are first and foremost deciding to build a large power station, and of course that decision would assume that this large power station, aimed at generating reliable baseload electricity, is actually required. We take that for granted. Why this need should be satisfied by proceeding to build CFR-1 at this stage should be clear from what several speakers have already said. The background that I envisage is something like this. We have to look forward to an energy economy which is very different from the one that we've all had for the last 25 or 30 years, because the hydrocarbons will simply not be there to sustain it. We have to decide how to generate this very large amount of energy demand towards the end of this century. We have heard that coal in this country can be an expanding resource, and we must all be glad. But it cannot expand to anything like the extent necessary to meet the full energy demand now carried by coal, oil and gas combined. I was trained as a chemist and increasingly I feel that oil and gas must be seen for what they really are; not materials to be burnt, but the most valuable chemical feed-stocks in the world. Men should use oil and gas increasingly to clothe themselves, to build their buildings, and even perhaps to eat, or feed to their animals; and not just to keep warm. We then come to what is left. What will fill the energy demand that will still be there? And it will still be there whatever sensible courses of energy conservation are put in hand immediately, because the industrial West will continue to use electricity for the uses that electricity are best suited. That is not only to keep one's houses warm, but also to use machines. Granted that sort of outlook, the basis of our submission on CFR-1 is that here we have a developed technology that would bid fair to put the energy supplies of this country and other countries, if it could be developed both safely and under conditions where bad international results did not follow,beyond doubt for a century or more through the great efficiency of the use of uranium fuel that a fast reactor economy promises.

CHAIR Should we do this even if it means that we haven't got enough money to invest in some other things, or in alternative fuel resources?

ROBERTS Well, 'some other things' is an extremely broad phrase. May I answer the second question concerning alternatives? When I was speaking earlier I said that at this stage investment in the fast reactor and investment in alternative sources—wind, waves, geothermal, even solar—are not in competition, because of the very different point of development the two have reached and the differences of unit scale. Towards the end of the century this country will be facing—fifteen years from now perhaps— a series of other big decisions in energy policy. By then, we would have hoped to have proved the fast reactor as one option and, of course, we may have failed to do so, but we will try. By that time, I hope, the possibilities of generating additional power from waves and wind will have been demonstrated, and their economics known, and the demand curve for power as a whole—for electricity as a function of the economy that we have at that time—will be better known. And from then on decisions will be taken on a strict basis of economics and of relative safety. But the basis for making those decisions doesn't seem to be here today.

COTTRELL We have been talking about the value of CFR-1 as an investment considered as the starting point for a series of fast breeder reactors to be built some years hence. Can you tell us about its value as a power station, without taking account of its potentialities for leading to other stations?

NICHOLSON The idea is thet CFR-1 would be operated for this experimental and learning process for a time, and when it had achieved a satisfactory performance in terms of reliability and everything else, perhaps it could then be used as a standard generator of electricity, and I would think its value at that time would probably be something like half the cost of what it would cost to build it.

COTTRELL So are you saying that discounted to the day when investment is put into the CFR-1, half of that initial investment can be got back in electricity return?

NICHOLSON No, I don't think that. I haven't discounted the capital cost spread over these many years that it is built. If you built it by 1986 and you've spent that amount of money, in 1991, say, when it's had its five years operation, its discounted value to the electricity producers should be then about half of the original capital cost of building.

WHYTE I want to ask Dr. Roberts whether his view on the investment was based on the projections that were put forward by the C.E.G.B. yesterday and have been put forward by others, of an increase in demand that will be double by the end of the century; and, if that is so, whether he would still have the same view of the value of the investment if, in fact, demand did not increase at these projections, and was not realised, or proved to be unreliable? If demand were held static to the end of the century would this still be a worthwhile investment?

ROBERTS The answer to your first question is No. My view was not based on that C.E.G.B. projection but on a number of comparative studies that we made at Harwell for the Department of Energy, published in a paper that was put forward for the Energy Conference in the summer. A number of different scenarios were put forward at that stage and the consensus view of a large number of possible futures was that a continuing reliance on a substantial nuclear programme was required. If you reach that point, I believe you should also go to a further point and take out an insurance policy against a world scramble for uranium.

CHAIR Now I would like to invite Mr. Cripps, Senior Research Officer at the Dept. of Applied Economics of Cambridge University and Advisor to Mr. Tony Benn. He has produced no written submission because I only invited him three days ago. Perhaps he would be willing to help us in these matters.

CRIPPS It's important to remember that the economic case and the fuel policy case for a programme of fast breeders can only be resolved very gradually over a long period yet to come, and nobody should really make up their minds on those grounds as early as this. It's also wise to remember that we don't have a fully-planned and centrally co-ordinated energy sector in this country, so the questions or the decisions about how fast to expand or whether to build a breeder programme at all will not in any meaningful sense be made by one individual but will be, to quite a large extent, determined by things which happen. There are two possible courses of future development. One would be that the economic case for the breeder programme gradually becomes stronger and stronger, if the electricity grid system on which it's predicated grows. At the moment it's suffering a set-back because of temporary competition from other fuels, chiefly gas. But if the grid system revives, and if the breeder looks like the cheapest way of expanding input to the grid, then there will be strong arguments on economic grounds for building a breeder programme. But we won't know about that for about another ten years, perhaps longer. The other eventuality is that the grid system continues to suffer competition from other fuels. I mean particularly the growth of solar energy. Possibly gas reserves in the North Sea will continue to be exploited and make a larger contribution than a lot of people think at the moment. Possibly all sorts of other forms of alternative sources of energy will gradully come into the market place. Possibly consumers will find it worth their while to instal a lot of insulation and to use energy much more efficiently. It's quite possible that the electricity system will not in fact experience the kind of growth of demand for a long time that it used to in the past.

Now if that happens, the case for plugging in an expensive programme of new investment in new generating capacity of any kind will be fairly weak. The case for the breaders will probably be at least as weak as that for additional coal-fired stations. So that's the kind of range of possibilities. But it's possible to make mistakes in this process. The big mistake is probably to commit ahead of time a lot of very expensive investment which then turns out not to be viable in market terms. I'm sure there's a very great anxiety among a lot of people that a breeder programme would really lock us into doing this. It's possible to imagine that we would build our 33 Gigawatts of breeder capacity, that they would turn out rather expensive compared to other solutions to energy problems which people will have found, that the electricity system will never run at full capacity, that it will have to charge very high prices or else receive heavy Government subsidies, and that a minority—or perhaps even a majority of energy consumers will pay the penalty, because for one reason or another they're locked into the electricity system in the same sort of manner that some people argue that all electric household consumers are paying a penalty at this moment due to over-expansion of electricity in the past. So that's the kind of economic problem which we might run into from a mistaken commitment or an early commitment which then turned out to have been false. It's not possible—for me at any rate—to really add much to that.

What are the presumptions now? I think it's too early to say. There is one presumption which has been correctly pointed out by Peter Chapman, which is that if we really have to depend on the grid for an awful lot of our energy in the year 2000 and beyond, and if we have to build large generating stations to feed the grid, and particularly if those generating stations are fast breeders, we shall have a very heavy investment requirement. It will take a long time for that investment to pay off because that is the nature of centralised generation and a grid system. If we have this very heavy investment requirement, whether or not we can cope with it is almost unknowable at the moment. If the British economy revives in the 80's, and if we become by contemporary standards a prosperous manufacturing nation in the 1990's, then putting up more central generating stations and extending the grid—or increasing its capacity—will seem a not impossible task. I don't think anyone would notice it very much. On the other hand, if we have continued economic problems of a substantial character, it will look an expensive sort of investment for the nation to undertake. If the present mood of the nation for strict economy in public expenditure and in heavy capital investments continues I don't think anybody will be interested in extending the power supply by this sort of means in the 1990's.

CHAIR I think I have a duty to put on behalf of this Hearing our main question. If we commit ourselves to CFR-1, will the Public Sector Borrowing Requirement be such that we will not be able to put the money that some people think should be put into the research and development of alternative supplies of energy?

CRIPPS I don't think it's possible to answer that, because the decision on public expenditure is much broader.

CHAIR So you can't answer it.

CRIPPS But it will be at the expense of something else, without any doubt.

CHAIR Would it be possible, for example, that we might not be able to spend so much money on rebuilding schools, or some other area of public spending?

CRIPPS Yes, I think it will certainly require some reduction in some other public spending.

ADAMS Are you implying that there's no need to make any decisions now?

CRIPPS I can only give a personal view. I don't think the case is proven or that it will make all that much difference to the economic benefits which the breeder programme might eventually give us, whether or not we decide within the next year or two to commit ourselves to CFR-1. But I haven't all that much confidence even in that view. I cannot see any compelling reason why it makes that much difference to decide now.

HENDERSON Here is a quotation from a paper submitted by Mr Surrey and Dr Dombey, which ends with the words, 'Licensing from abroad might well be the best method of acquiring F.B.R. technology, rather than building CFR-1.' Would you like to comment?

CRIPPS I'm aware of considerations on both sides, but I have difficulty in reaching a view because they work in opposite directions. One argument is that if we do not sustain our capacities and capabilities in this area by public expenditure, then the CFR-1 would be the most suitable form; we will simply not have the organisation, the manufacturing know-how, to make a major contribution to an international process of manufacturing in the future.

COTTRELL. Do you not think that there is value—when that ten year period of which you have spoken has arrived—in knowing that we have proved ourselves able to construct and operate a successful commercial reactor?

CRIPPS I put the argument in favour of that view a moment ago. The argument against it is that we would run a risk of committing ourselves to a special and particular U.K. technology and to have most of our resources tied up in that, and that we might—by not doing anything now—have a much better opportunity if we went flexibly, as sub-contractors, into international manufacturing programmes of the 1990s. That is, if we hadn't attempted to build a whole system of our own in the interim. I don't know which argument in the end would win.

---

[1] Culham Study Group Report on Fusion Reactors and the Environment. R. Carruthers *et al,* Culham Laboratory Report CLM-R 148, June 1975

Fusion Power and the Environment, F. N. Flakus, "Atomic Energy Review", Vol. 13, No. 3, pp. 588-614, 1975, International Atomic Energy Agency, Vienna.

[2] A 1300 MWe CFR-1 with 80% load factor and 10% distribution losses would provide 0.7 million tons oil equivalent, perhaps in 1988-1990. Several middle of the road scenarios show that after 1990 the U.K. must find each year replacements for oil and gas, either through new supplies or savings, of around 17 MTOE per year.

[3] Robert Williams and Marc Ross, 'Energy in a Sustainable Society' (draft book manuscript, August 1976): Princeton University.

[4] There are safer alternatives to plutonium breeding: for example, the use of Thorium-232 to breed Uranium-233, which can be mixed with natural U-238 to render it virtually useless as a bomb material. Slower nuclear growths plus new uranium ore discoveries and better production economics could delay for many decades the need to breed nuclear fuels.

[5] An apology was later received.

# *Chapter VI*

## Epilogue

### *by the Chairman*

I am instructed to end with an epilogue. I'm not going to pass a verdict on CFR-1 because I've repeatedly said that our aim on this Panel is not to pass a verdict but to expose arguments and assumptions. Nor will this be an epilogue in the sense in which the word is often used in the T.V. networks—comforting words from a clergyman before turning in for the night. I want to do three things. First, I want to explain why the British Council of Churches invited me to arrange these Hearings, secondly to pass a few theological reflections upon them, and finally to say something about the nature of the Hearings.

First, the British Council of Churches did not ask me to undertake this job because the Churches in this country have contracted out of religion and have gone into whatever particular secular subjects happen to occupy popular interest at the moment. Far from it. Christianity is, by its nature, a world-oriented religion. At this very season we are preparing for Christmas, which Christians believe commemorates a divine irruption into this world; and if God took it as seriously as that, we can do no less. In other words, the Council's interest in our future energy supplies is not an aberration from its main concerns, but issues directly from its deepest convictions about the nature and activity of God. Moreover, the Churches are not merely concerned with the condition of individuals in the world but with the decisions of society as a whole, especially as more and more decisions affecting people's lives are taken on a larger and larger scale. This public hearing is part of the Council's attempt to minister to society, and in so doing we wish—and I hope we have—co-operated with all men of good will. For this hearing we have wanted, and we have been most fortunate to get, the best brains on this subject in the country.

Now for a few theological reflections. It is sometimes thought that Christianity is about the subjective religious experiences of individuals. In fact, it is part of Christian conviction that the whole world belongs to God, that the world has a point and goal of its own among the varying goals of this expanding universe, in which our planet occupies so insignificant a place. Human beings dare not place in jeopardy the future of this planet. They have a duty to leave the world no worse than when they find it. Indeed, if they are co-workers with God they have a duty to embellish the planet, and to restore those parts of it which they have allowed to degenerate. Man has a genuine concern and respect for nature's inherent values. If he is made in the image of God, he must behave in a God-like or responsible manner.

We are, however, human beings, and people matter most. If we use resources today, we must also take thought for people tomorrow. We must not deprive posterity of the options to lead fully human lives. Of course we must take risks—that is part of what it means to be human, to take risks, but responsible risks; risks that will not deprive posterity of its birthright. And so the overriding concern of everybody, on whichever side of the fence they might be, has been for safety; and very proper that is. It is precisely because of our concern for posterity that we have taken thought for the year 2000, the year 2005, and 2010, and thereafter. The lead-time required means that something must be decided now about what will happen then.

Here I mention a third principle, on which I hope too we will all agree; that we have an obligation to plan so that human beings can develop as fully as possible their humanity. Few would doubt that a rising standard of material living can greatly help when it lifts people out of poverty, although Christians can never forget the Beatitude which begins: "Blessed are the poor." The question that we need to ask, and I wish it could have been asked more in this enquiry, is whether or not the increasing supplies of energy which we could expect from a nuclear economy would involve us in a kind of society in which we will become more fully human, or whether we would find that we would become dehumanized by consumerism and by our increasing distance from our natural human state. So the question arises whether a nuclear economy would help or would hinder people from becoming more fully human. And we must ask whether a nuclear economy in these islands would be a help to developing countries as well as to the developed countries, for we are all brothers and sisters on the planet. Here I hope we would agree on the principle, but it is clear from what we have heard that we disagree about its application.

Well, I put before you these three theological reflections; one concerning the planet, one concerning posterity, and one on what it means to be fully human—men and women made in God's image. And I suggest that these three considerations—I believe they are theological in origin—underlay all our Hearings. I add a final theological reflection: it is this. Few historians, looking back over the span of history, would say that many countries have been able to control their own destinies. We must of course plan, we must be rational, we must be prudent; but so often men and countries are at the mercy of circumstances. Today we are at the end of an age; indeed, I would say that we are at the end of an epoch. Just as men passed from the nomadic to the agricultural epoch, so it seems we are passing from the industrial to the electronic and nuclear epoch, and it is likely to be a turbulent transition. Stability is likely to be fragile. Theologically speaking, we would express this by saying that all things are in the providence of God, and nothing—not even catastrophe—can prevent the final fulfilment of His purposes, though it may delay them a few million years. We must accept that the unexpected often happens. We may plan for growth; we may be over-ruled by events.

I pass now briefly to the purpose of these Hearings. I believe that they are almost unique in this country although I am ready—I hope always— to be corrected. We have managed to gather great good will, so that within five weeks the leading experts in our nation have willingly co-operated in speaking here openly, publicly, and most good-humouredly about crucial national issues which deeply divide people, and they have done so in a language and with a simplicity which enables ordinary people to understand. Modern democracy no longer means that we can delegate all decisions straight to our Parliamentary or our Borough Council representatives. If we are to be fully human we all need some share in the decision-making process, especially in major and crucial issues of policy; and all national issues of high policy have an ethical aspect. But we are not in a position to make ethical judgements unless we know something of the economic and technical aspects of these matters. And I believe we have been given them. I have been asked whether I shall make a formal submission to the Secretary of State for Energy as he seemed to imply. I shall not. There will be no verdict. I have been asked whether there will be a Statement of Concern issued, similar to that of the Church of Scotland. I don't know. But if it is, this Panel will not have anything to do with it. It would be a matter for the British Council of Churches.

I believe that we have managed to ventilate the underlying issues on a very broad front—goals and needs, risks, feasibilities, alternatives, investments—in two days. As I say, I'm deeply grateful for those who have co-operated, and I apologise for any deficiencies in my Chairmanship. If there have been matters which have not been as fair as they should have been, I apologise. I have tried to make them so. I hope that we have been setting a precedent by holding these Hearings. I would like to think that there will be many many more such Hearings; and I am proud that the first initiative has been taken by the British Council of Churches.